Fluid Frontiers

Fluid Frontiers:
New Currents in Marine Environmental History

Edited by

John Gillis and Franziska Torma

The White Horse Press

The White Horse Press, 10 High Street, Knapwell, Cambridge, CB23 4NR, UK

Set in 11 point Adobe Garamond Pro
Printed by Lightning Source

British Library Cataloguing in Publication Data
A catalogue record for this book is available from the British Library

ISBN 978-1-874267-86-7 (HB)

CONTENTS

ACKNOWLEDGEMENTS

This book originated in a conference, *Fluid Frontiers: Exploring Oceans, Islands, and Coastal Environments*, sponsored by the Rachel Carson Center for Environment and Society. A call for papers went out from Munich, directed to junior scholars around the world. It resulted in seventeen people from four continents being invited to Rockland, Maine, in October 2011. The Island Institute and its president, Philip Conkling, agreed to host our three-day gathering, organised around discussion of written papers and field trips to Rockland's marvellous Farnsworth Museum and the working waterfronts of Rockland and Port Clyde, which included a visit to the latter's fishing cooperative, Fresh Catch. Philip Conkling opened the conference with a talk entitled 'Maine Islands in Global Perspective'.

In what proved to be an extremely productive format, we invited six senior environmental historians to join us as commentators and mentors. Philip Steinberg, then at Florida State University (now at University of Durham) lent his expertise in oceanic geographies. A Pacific perspective was provided by Paul D'Arcy of Australian National University. Michael Pearson, from the University of Technology in Sydney, brought his knowledge of littorals to bear, while Steve Mentz of St. John's University in New York added a literary dimension to our discussions. The history of fish and fisheries was ably represented by W. Jeffrey Bolster of the University of New Hampshire, while Helen Rozwadowski of the University of Connecticut at Avery Point contributed her immense knowledge of the oceanographic and marine sciences.

Although their papers do not appear in this collection, Kimberly Peters, then at Sheffield University; Julia Heunemann of the Bauhaus University-Weimar; Robert Deal of Marshall University; Ryan Jones of Idaho State University; and Samuel Seneyon Olaoluwa of the Osan State University in Nigeria were all active participants in every dimension of the meeting. Also in attendance was Ingo Heidbrink, Old Dominion University Norfolk, Virginia.

We are particularly indebted to Christof Mauch and Helmuth Trischler, directors of the Rachel Carson Center for Environment and Society, who took a personal interest in the project and made many suggestions, including the recommendation of The White Horse Press, that have been invaluable.

Members of the Rachel Carson Center's staff, including Angelika Moeller, Lisa Spindler, Martin Spenger, Marcus Andreas and Pierre Lipperheide helped with the organisation of the conference. Nancy and Sally Perkins of the Island Institute also deserve thanks for all they did on and off the conference site. The conference poster – the cover of the present volume – was brilliantly illustrated by Tom Killion, who draws his inspiration from the California coast at Point Reyes, just north of San Francisco.

Our collaboration as organisers and editors spanned not only continents but generations. The internet made things so much easier, but it turned out that we were particularly compatible, intellectually and temperamentally. Our jointly-authored introduction is meant to guide readers to what is a rapidly developing field, one that is full of opportunities for those who wish to be on the frontiers of the new environmental history.

John R. Gillis
Professor Emeritus, Rutgers University
(now resident in Berkeley, California)

Franziska Torma
University of Augsburg

September 2014

To the Memory of Rachel Carson,
Scientist, Poet and Truth Teller

Introduction
Fluid Frontiers: New Currents in Marine Environmental History

John R. Gillis and Franziska Torma

AMPHIBIOUS HISTORY

Fluid Frontiers: New Currents in Marine Environmental History represents a substantial step toward closing what we have come to think of as the blue hole in environmental history, a strangely landlocked discipline which, to date, has largely neglected the seventy per cent of the earth's surface covered by water. This collection dissolves the line between land and sea. It suggests the many ways that we might move toward an amphibious history. Water is the source of all life, and humans are like all other species in their dependence on fresh and salt aqueous environments. Human modes of amphibiousness have varied widely over time and space. Some peoples have depended almost entirely on the sea but no culture, even the so-called Sea Gypsies, has ever abandoned land entirely. Even those who live far inland and have never seen the sea are exposed to its effects on temperature and weather. Even the most landlocked peoples have an image of the sea that affects the way they interpret the world.

We are more informed about some amphibious human encounters than others. Maritime history, which until recently has been largely about ships on the surfaces of the deep seas, is well established. We are far less acquainted with lake and riverine peoples, and those who occupy swamps and estuaries. There are the great classics of sea literature from the Odyssey to *Moby Dick*. Poetry, art and music are all intensely watery. The history of certain marine species has attracted more attention than others, with whaling being the most thoroughly documented (Creighton 1995; Burnett 2007). Science started to research the smaller fauna down to the level of microorganisms in the second half of the nineteenth century. The humanities have been getting around to researching the meaning of these organisms, all vital to the health of the oceans, only in very recent times.

Land has never been separate from water. Islands are defined by it, and continents are no less affected by the oceans that surround them. The development of *Homo sapiens* can be traced back to the moment when humans came down to the sea on the southern tip of Africa, some 164,000 years ago. Since that time, coasts have been an exceptionally productive home for our species. Today, they provide leisure as well as employment, nourishing the soul as well as the body in innumerable ways. In this century, humans have become more rather than less amphibious, learning to swim, surf and snorkel, becoming efficient submariners for the first time in history (Corbin 1994). During the Cold War, Conshelf and other sealab programmes enabled researchers to live on the sea floor for periods of time, giving rise to speculation about the colonisation of the ocean bottom. The French oceanographer Jacques-Yves Cousteau even dreamed of an age of 'mer people', hybrid human-machine creatures adjusted to maritime environments (Kroll 2008, pp. 152–188).

We have not matched the once landlubberly whale in its ability to return to water, but we must also be aware of animals like the sea turtles and elephant seals who depend on land to birth and nurture their young. Today many species arrive on land through research stations, marine laboratories, aquaria and zoos, enriching our worlds scientifically and aesthetically. In effect, much of the recent discovery of the sea has taken place on land or from the land.

The discovery of the seas in the fifteenth and sixteenth century was crucial to the exploration of lands and peoples previously unknown to Europeans. The Enlightenment was largely a waterborne enterprise (Bentley 2007). The travels of the enlightened Era opened up the Pacific islands that were to become the focus of scientific anthropology and, later, biogeography (Kuklick 1996, pp. 611–638). As Richard Grove has shown, Europeans discovered the fragility of ecosystems on colonial islands where the introduction of alien plants and species had such devastating effects in the Early Modern period. The notion of 'ecology' developed by Ernst Haeckel in the second half of the nineteenth century was a product of his oceanic explorations over decades. In observing there the processes of exchange and adaptation by aquatic creatures, he came to think of nature as a unified dynamic system (Grove 1996). We shall see in this volume how oceans and islands contributed to the ecological thinking of Alexander von

Humboldt and to the evolutionary theories of Charles Darwin and Alfred Russel Wallace.

Our contributors show that the distinction between eternal sea and historical lands does not hold. The line between land and water dissolves when we consider those places – wetlands, estuaries, shores – where these elements intermix and interact. Rather than drawing lines, we prefer to dissolve them in favour of more fluid categories that better represent the amphibious nature of the worlds we inhabit. Here we are concerned as much with brown as blue waters.

TRIDIMENSIONAL HISTORY

There is a third dimension too often ignored by historians and geographers. The vertical has too often been subordinated to the horizontal, especially when it comes to the oceans. Western sciences had long known a great deal about the heavens above before getting around to investigating the waters below. Captain Cook sailed in 1769 to observe the transit of Venus, but in the course of his journeys opened new horizons in the little-known Pacific (Richardson 2007). But it was not until 1853 that Matthew F. Maury organised the first international maritime conference in Brussels. Two years later, he published in the name of navigational safety his *Physical Geography of the Sea*, which charted winds and currents (Maury 1855).

Much of the investigation of the undersea was taken up in the interests of shipping and fishing, but in 1860 the Mediterranean telegraph cable was raised for repair from the depth of about 1,000 fathoms. A yet unknown microorganism attached to it caused amazement among both the public and scientific circles, for it refuted Edward Forbes' widely accepted azoic theory, which theorised that life could not exist below 300 fathoms. In the next half century, the third dimension was explored by the British *Challenger* expedition (1873–6), the United States *Albatross* (1887–89), and the German *Valdivia* (1898). Nearer to shore, further insight was provided by dredging operations, fathoming and the commercial reports of whalers and fishers.

As Helen Rozwadowski has shown, the deep sea was thought of in terrestrial terms as the last 'undiscovered continent', using conceptions drawn from land-based geography and geology (Rozwadowski 2005, pp. 37–65). In time, the complexity of the ocean bottom, its diverse layers as well as its multiple fauna, came into focus. With deep-sea sounding, the sea took on volume for the first time. Its resident species, whose horizontal movements had been partially known earlier, were now seen have a vertical dimension as well. The ocean was shown to be, like mountains, divided into zones, confirming Alexander von Humboldt's early speculations on the vertical as well as horizontal dimensions of all ecologies. From these new perspectives, the oceans came to be understood on their own terms, no longer as static land-like territories but as highly dynamic fluid systems with a life all their own.

Today, oceans compete with space as a scientific frontier (Rozwadowski 2012, pp. 578–602). And it is useful to think of them in this way because frontiers are associated with movement and with unique human cultures adapted to those conditions. Normally we think of frontiers on a horizontal plane, but we can also think of them in the third dimension, reaping a harvest of new insights, not just about the fish and animals of the oceans but about ourselves as an amphibious species. In this volume, we encounter several sea frontiers – extractive frontiers constituted by various fisheries, leisure frontiers created by the modern tourist industry, imperial frontiers established by naval powers. Frontier societies have always been different from settled ones and maritime frontiers have their own peculiarities, their own ecologies, which are explored here.

Very often the notion of frontier conjures up the vision of empty lands just beyond the horizon. For a very long time, the sea shared this vacancy. But this is no longer viable, for two hundred years of exploration of ocean depth as well as breadth has wholly refuted the idea, inherited from the ancients, of the ocean as void. We now know it to be not empty but full, not just of life, but of meaning. The horizon of our imagination is now deep as well as wide. Since Jules Verne, the submarine world has been one of the fecund sources of literary and filmic inspiration. To the inventory of material riches the sea provides must be added its aesthetic treasures. Our world is immensely richer for the third dimension of this fluid frontier.

SEA AS HISTORY

In this collection the sea ceases to be a stage on which history is acted out and becomes instead a historical actor in its own right. It is not surprising that the notion that 'the sea is history' comes more easily to islanders like Derek

Walcott and Epile Hau'oufa (Walcott 1930; Hau'ofa 1994). As Jonathan Raban has put it, 'people who live on continents get into the habit of regarding the ocean as journey's end ... Islanders also know that how the sea goes on and on, a continuous loop of shoreline and life, without terminus' (Raban 1988, pp. 300–301). Our landlocked narratives are a legacy of nineteenth century historiography, which chronicled and mapped territory in exquisite detail, but left that which lay offshore largely blank. Maritime and fishery historians will object that they already have an established historiography, but it is important to note that they write about what takes place *on* rather than *in* the sea. Only very recently have fish and marine animals been treated as historical subjects (Bolster 2008, pp. 19–47).

One of the most powerful tropes of western continental culture is that of the so-called 'eternal sea'. It has its roots in ancient history, but also modern origins in nineteenth century romanticism, which juxtaposed nature to history, sea to civilisation. For Henry David Thoreau 'the ocean is a wilderness ... wilder than a Bengal jungle and full of monsters' (Thoreau 2004, p. 112). In western cultures, the sea was viewed as a trackless void, resisting human agency. Even Rachel Carson called it the incorruptible sea, denying human impact. Only recently have ecologists made a place for man in nature, but primarily on land. We are only beginning to appreciate what enormous effects humans have had on the sea, and the sea on them (Roberts 2007).

Here we provide multiple examples of how to render the sea as history. Though we are not unconcerned here with deep time, is important to note that water is earth's original element. Only a profound understanding of millions of years of geological, microbiological and climate change can fully explain the oceans as they exist today. In this planet's vast, interconnected ecological system, all the parts must be understood in terms of all the others. The sea affects the land, not just at the shore but as a global thermostat. In turn, seas are altered by deforestation and the resulting siltation, by carbon emissions and acidification.

Then there are also the long timelines of marine plants and animals, whose evolutionary paths intersect with our own. We must now add *homo sapiens* to the inventory of marine amphibians. We, like all other life forms, came from the sea (Carson 1951). Humans depend on it, but they have never been able to exist apart from land. Even the so-called 'sea people' of South-east Asia return periodically to shore; and those who make their lives *from* the sea do not live *in* it. As Michael Pearson rightly insists, there is no such thing as a separate maritime world. Mariners and fishers have never been a race apart. Coasts are always porous frontiers, along and across which flow not only marine but human species. It is this quality of movement that sets the littoral, a true ecotone, apart from interiors.

In this volume, history does not begin or end at the edge of the sea. It encompasses the currents of human and natural history which, to this point, have been separated, combining them into one continuous interactive flow. The ocean becomes a historical actor here, but only one of many. One might question whether the sea makes history as we do, conscious of itself as an agent of change. But human actions are not always intentional either. What we can say is that both operate in an ecological force-field which honours no line between land and water.

One of the unique features of this collection is the way it brings culture into history. Helen Rozwadowki emphasises the various epistemologies – scientific, aesthetic, philosophical, literary – that have brought us to our current understanding of the sea and continue to push us to new discoveries. Millions know the sea through art and fiction who have never actually experienced it physically. Until recently, oceans have occupied a larger part in literature than in science. Historians have also been tardy in making their contributions, but now they too are granting the sea the importance it deserves. This latest discovery of the sea is clearly temporal as well as spatial. As Stefan Helmreich tells us, things like waves, which were previously thought of as a natural phenomenon, can be shown to have a history.

KNOWING THE SEA IN NEW WAYS

Knowledge of the oceans is currently in healthy flux. New discoveries have challenged conventional wisdoms. New technologies, ranging from satellite imaging to the use of DNA traces, have expanded so quickly the range of what we can know that much has not yet been assimilated by the scientists themselves, far less the general public. When this happens it is tempting to question the validity of science itself, especially when the new findings clash with vested interests, as has been the case in recent climate science controversies. But a certain degree of uncertainty is a useful prompt to further investigation, and what is certain is that the marine sciences emerged recently compared to astronomy and geology.

The same must be said of the marine-related humanities and social sciences. The systematic study of ocean geography is just beginning; as this volume will demonstrate, sea literature, one of the world's oldest genres, has only begun to be subjected to sophisticated criticism. The sociology and anthropology of seafaring peoples is also entering a new phase, embracing a raft of aquatic activities – surfing, snorkelling, and swimming – that have yet to have systematic treatment. Scientists are discovering that, despite advances in technology, they are not always equipped to understand or express what they find on fluid frontiers. The leader of the first German deep sea expedition on the *Valdivia*, Carl Chun, found that he could not express in words the marvels that he found in the Malaysia's coral reefs (Chun 1900, p. 319). Colours, sounds and smells are not easily objectified. In addition to verifiable knowledge, oceangoing travellers brought back accounts of wonders and unexplained mysteries. Indeed, the lure of the unknown, even the unknowable, continues to be a motive for exploration.

Science has always been driven by human emotion and desire, though only recently has it been possible to fully acknowledge this. In the 1930s, the French philosopher Gaston Bachelard began to integrate imagination into his phenomenological studies. Inspired by Jungian psychology, he bravely took on water as subject. Its apparently ephemeral, unsubstantial nature did not deter him. Working in what is now a recognisably postmodern manner, he approached his subject not as a natural object but as a cultural construction (Bachelard 1993, p. 183). Several of the papers in this volume follow in his footsteps, bringing to bear linguistic and other methods to explore the meaning as well as the substance of marine phenomena.

The sources available to the study of fluid frontiers range from scientific accounts to travel writing, diaries, photographs, legal documents, fictions of all kinds, maps. To know the oceans in a comprehensive way demands trespassing the borders of many disciplines. In our volume, the humanities and sciences are co-constructors of new knowledge. They depend on one another for illuminating metaphors and images. We foreground the role that imagination has played in emergence of the environmental sciences.

NECESSITY OF MARGINS

Until recently, oceans have been marginal to both history and geography, and to the environmental sciences more generally. Green environments have had much more attention than blue ones, though the latter account for seven tenths of the globe. Historians set temporal beginnings and endings at the edge of the sea. Geography has been equally terrestrial. But, as the essays in this volume suggest, things are now changing. That which was marginal – swamps, coral reefs, coasts, waves, estuaries – is now front and centre. Even more important, marginality itself has gained new significance. As Wendell Berry has argued, our understanding of environmental change will be enhanced if we consider margins: 'If change is to come, then it will have to come from the outside. It will have to come from the margins' (Berry 1996).

One of the intentions of this volume is to disassociate margins from marginality, to demonstrate their centrality to ecological understanding as such. Like Timothy Morton, we mean to 'support these margins. As a matter of urgency we just *cannot* go on thinking of them as 'in-between'. We must choose to include them on this side of human social practices, to factor them in to our political decisions.' (Mason 2009) In the modern era, margins have often been marginalised to the point of their extinction. Over the last two centuries, wetlands and estuaries have filled in, swamps drained, corals blighted and shores engineered to the point they have lost their original features. On the land, industrial agriculture has been ploughed to the very edge of field, destroying the margins that not only provide living space for birds and animals, but prevent disasters produced when monocultures supplant biodiversity (Berry 1996, pp. 179-180).

Coasts provide a particularly good example of what happens when a broad margin is reduced to a sharp edge. Before the eighteenth century, the shore was not yet understood as a *coast line*, clearly dividing land and water. Coasts, as understood by mariners, extended well inland and encompassed offshore islands. Sea charts marked landmarks and harbours, but drew no line where we would now expect to find the coast. To the pre-modern world, coasts were what Rachel Carson would describe as a marginal world, an 'an elusive and indefinable boundary' (Carson 1998, p. 1). This was not a place 'in between', not the edge of something else, but a liminal place belonging neither entirely to land not to sea, inhabited by amphibious species, including our own (Gillis 2012).

Today we call such a place an *ecotone*, somewhere that two or more ecosystems come together. Shores, when left in their natural state, are a porous zone whose dimensions are constantly changing, a fluid frontier constituted partly of land, partly of water, often fresh as well as salt. These are among the biologically richest places on earth, where, for millennia, humankind has made one of its most productive homes. Things began to change when, in the course of the nineteenth century, shores were surveyed, mapped and ultimately engineered to separate land and water. In the process, what had once been a living, moving thing was pinned down to anthropocentric specifications. Coasts lost the quality of a margin and became an edge. In the process, they lost their hybridity, biodiversity and ability to renew themselves through the natural actions of the tide and wind.

The difference between an edge and a margin is poorly understood. Edges are human constructs, lines drawn where none existed in nature. By contrast, margins are fractal, the product of natural processes. Edges are points where things start and stop; margins are where things flow. Edges separate; margins connect. As Edward Casey puts it, the margin is 'embracing and open, in contrast with an edge, which emphasises alterity and difference' (Casey 2007, p. 69).

At the edge, time's arrow points in one direction; in margins, time is more cyclical. Edges belong to the world of linearity, margins to the world of circularity. An edge is by definition something 'in-between', without a history or geography to call its own. It always belongs to something else. It is a momentary transit point on the way to elsewhere. By contrast, a margin is not only a place in its own right, but has a rhythm all its own. You can be *on* an edge, but never *in* one. A margin can be narrow or wide, but it is capacious. Margins are habitable, edges are not.

This collection re-centres our attention from the deep blue sea to the brown waters of the near shore. Its emphasis is on the ecological continuum that is land and sea. It is concerned with amphibious species, including our own, who lead a terreaqueous existence. Now that we live in what Zygmunt Bauman calls a state of political and social fluidity, not only borders but edges are crumbling (Bauman 2000). It is time that we embrace rather than resist margins, appreciating the part they have played in evolution and how necessary they are to the survival of the planet.

DISCOVERING THE SEA

The sea has been discovered many times over. In the ancient world of the West, it was conceived of as a river, *Oceanus*, encircling an earth island, *Orbis*

Terrarum. In the fifteenth century it was re-imagined to be several bodies of water, but just how many has never been entirely clear. Since the term Seven Seas was first used in ancient times, the bodies of water constituting them have changed many times over. Names have also changed with bewildering frequency. The term 'Atlantic' was not used as it is now until the mid-sixteenth century and notion of a Southern Ocean did not disappear until after World War II. Today, the seven seas have been reduced to three and, in this global era, the notion of one great continue ocean is gaining ground. With seventy per cent of the surface covered by water, Planet Earth could be more accurately described as Planet Ocean (Lewis 1999, pp. 188–211).

Oceans are constantly changing shape and volume. They do so daily, monthly, and have done so over eons. In this period of sea level rise, the shifts are even more dramatic. It is startling to think that until quite recently seas were seen as unchanging. Oceans were understood to be the oldest part of the earth. While their surfaces could be agitated by winds, the deeps were considered dead, incapable of sustaining any kind of life (Kunzig 1999, prologue; Hamilton-Paterson 2009, chapter V). Today, we know that the entire sea sustains life at all its levels, with huge storms below as well as above its surfaces.

As the essays in this volume illustrate, the sea is alive, perhaps the most animate element in our environment. Unlike other accounts, which treat only ocean surfaces, our approach is three-dimensional. It not only probes the depths, but recognises that bottoms as well as shores are part of the terraqueous environment. As Michael Reidy insists, the vertical as well as horizontal must be considered in any true ecological approach. In this most recent discovery of the sea, oceanic environments have revealed themselves to be much more dynamic and complex than previously recognised. Here we have deliberately muddied the waters by bringing land, water and air together, refusing the binaries that have previously limited our understanding.

The oceans are now seen from so many different perspectives that no sea is ever quite like any other, far less just one stable object over time and space. The sea is a different thing to seafarers than it is to landlubbers. Pacific islanders experience it in a very different way from their Atlantic counterparts. The sea perceived by eye is different from that heard by ear or sensed by touch. Joseph Conrad worried that modern steamships would erase eons of experience. 'The machinery, the steel, the fire, have stepped in between man and the sea', he wrote in 1924 (Conrad 1924, p. 7). But he need not have worried. For never has humanity been so intimate with the oceans as today, when throughout the world there is a surge to the sea, with half of all the global population living within a hundred miles of a coast (Helvarg 2001). Few any longer make their livings there, but coastal populations are growing at an astounding rate. As fish stocks collapse, the shore has become primarily a place of leisure rather than work, yet we identify with the sea as with no other aspect of nature. It infuses our language, our art, music and literature (Mack 2011). With the popularity of swimming, surfing, snorkelling, Rachel Carson's prophecy that *homo sapiens* would ultimately return to the sea 'mentally and imaginatively' (Carson 1951, p. 15) will have been sustained.

'We love the ocean, we use the ocean, but we don't think enough about the ocean', writes David Helvarg (Helvarg 2001, p. 245). The challenge now is to protect it from being loved to death, building a sustainable relationship, understanding that we belong to the same ecological world, dependent on it as it is on us. We hope that this volume will not just contribute to our knowledge of the sea but lend support to all those efforts to save it to the extent still possible. If this most recent discovery of the sea is successful in that respect, then we will have contributed something very important.

September 2014.

BIBILIOGRAPHY

Bachelard, G. 1993. *Waters and Dreams: An Essay on the Imagination of Matter*. Dallas: The Dallas Institute of Humanities and Culture.

Bauman, Z. 2000. *Liquid Modernity*. Oxford: Polity Press.

Bentley, J., R. Bridenthal and K. Wigen. 2007. *Seascapes, Maritime Histories, Littoral Cultures, and Transoceanic Exchanges*. Honolulu: University of Hawaii Press.

Berry, W. 1996. *The Unsettling of American Culture and Agriculture*. San Francisco: Sierra Club Books.

Bolster, J. 2008. 'Putting the Ocean in Atlantic History: Maritime Communities and Marine Ecology in the Northern Atlantic, 1500–1800'. *American Historical Review* **113** (1): 19–47.

Burnett, D. Graham. 2007. *Trying Leviathan: The Nineteenth Century Court Case that Put the Whale on Trial and Challenged the Order of Nature*. Princeton: Princeton University Press.

Carson, R. 1951. *The Sea Around Us*. New York: Oxford University Press.

1998. *The Edge of the Sea*. Boston: Houghton Mifflin Company.

Casey, E.S. 2007. 'Borders and Boundaries: Edging into the Environment', in S. Cataldi and W. Heinrich (eds) *Merleau-Ponty and Environmental Philiosophy: Dwelling on the Landscapes of Thought.* Albany: State University of New York Press. pp. 67–92.

Chun, C. 1900. *Aus den Tiefen des Weltmeres.* Jena: G. Fischer.

Conrad, J. 1924. *The Mirror of the Sea.* Garden City: Doubleday.

Corbin, A. 1994. *The Lure of the Sea: Discovery of the Seaside in the Western World, 1750–1840.* Berkeley: University of California Press.

Creighton, M.S. 1995. *Rites and Passages: The Experience of American Whaling, 1830–1870.* Cambridge: Cambridge University Press.

Gillis, J.R. 2012. *The Human Shore: Seacoasts in History.* Chicago: University of Chicago Press.

Grove, R.H. 1996. *Colonial Expansion, Tropical Island Edens, and the Origins of Environmentalism, 1600–1860.* Cambridge: Cambridge University Press.

Hau'ofa, E. 1994. 'Our Sea of Islands'. *The Contemporary Pacific: A Journal of Island Affairs* **6** (1): 148–58.

Helvarg, D. 2001. *Blue Frontier: Saving America's Living Seas.* New York: W.H. Freeman and Company.

Hamilton Paterson, J. 2009. *Seven Tenths: The Sea and its Thresholds.* New York: Europa.

Kroll, G. 2008. *America's Ocean Wilderness: A Cultural History of Twentieth Century Exploration.* Lawrence: University of Kansas Press.

Kuklik, H. 1996. 'Islands of the Pacific: Darwinian Biogeography and British Anthropology'. *American Ethnologist* **23** (2): 611–38.

Kunzig, R. 1999. *The Restless Sea: Exploring the World Beneath the Sea.* New York: W.W. Norton, prologue.

Lewis, M. 1998. 'Dividing the Ocean'. *Geographical Review* **89**: 188–211.

Mack, J. 2011. *The Sea: A Cultural History.* London: Reaktion Books.

Mason, T. 2007. *Ecology without Nature: Rethinking Environmental Aesthetics.* Cambridge: Harvard University Press.

Maury, M.F. 1855. *The Physical Geography of the Sea and its Meterology.* New York: Harper & Brothers.

Raban, J. 1998. *Coasting: A Private Voyage.* New York: Penguin.

Richardson, B.W. 2007. *Longitude and Empire: How Captain Cook's Voyages Changed the World.* Vancouver: University of British Columbia Press.

Roberts, C. 2007. *Unnatural History of the Sea.* Washington, D.C.: Island Press.

Rozwadowki, H.M. 2012. 'Arthur C. Clarke and the Limitations of Oceans as Frontier'. *Environmental History* 17 (3): 578–602.

2005. *Fathoming the Ocean: The Discovery and the Exploration of the Deep Sea.* Cambridge: Harvard University Press.

Thoreau, H.D. 2004. *Cape Cod.* New York: Dover Books.

Walcott, D. 1930. 'The Sea as History'.

I. REIMAGINING OCEANS

~1~

The Swamp: Waterland of the Semi-Wild

Adam Keul

INTRODUCTION

Since the middle of the twentieth century, we have witnessed the decline of the swamp. Though the material context of this fluid frontier between social land-space and ecological nature-space still covers thousands of acres of coastal peripheries, by depicting the swamp as a wetland and conceptually over-broadening it, ecologists and historians alike have reified its decline. Rather than being treated as cultural spaces, swamps have been formulaically lumped into the wetland category and discursively 'made pure' (or dehumanised) as a political tactic to help illustrate their downfall from the pristine. Of course, this was never the intention. Perhaps the most poignant calls for environmental preservation in the twentieth century have come from authors who seek to shed light on the treacheries of destruction through development. However, we have reached a point where appreciation for these environments is paradigmatically bound to understanding their *ecological* downfall. I argue that reliance on narratives of decline may bring attention to threatened natures, but does so by ignoring the lived cultures of places.

Cronon (1992) seems to foresee this problem and offers some suggestions on ways to diversify this project in environmental history. As he puts it, 'nature and the universe do not tell stories; we do', and imposing narratives on landscapes that truly are in a constant flux is a dangerous presupposition (1368). Environmental histories often lean on the assumption of a holistic spatial ecology, a focus which denies the diversity of ways people experience the environment. Ellis also poses this critique. 'Although the intention behind this silver-bullet approach to understanding the global environmental crisis has been to provide the environmental movement with a clear focus and agenda, its impact has been very nearly just the opposite' (Ellis 1996, p. 266). Cronon (1992) calls for several changes in methodology that can counteract the uniform representation of change as a downward spiral, or at least produce 'good histories' of this process. One still valuable suggestion was his call to include a variety of different voices that have yet to speak (or be spoken for).

Motivated by such calls for a paradigm shift, I suggest that we bring the swamp back to life. Through research into discourse and participant observation of the 'live' production of swamp narratives in the context of boat tours, I found that Louisiana's Atchafalaya Basin swamp is riddled with stories of lost and declining nature, yet is concurrently appreciated (both economically and aesthetically) by recognising hybridities of nature and culture. By fashioning the concept of 'semi-wild', swamp tour guides, who make a significant impact on people's understanding of this place, have blended tropes of natural purity with histories of human–swamp interaction, both sustainable and damaging. Applying semi-wildness gives local swamp tour guides a narrative of environmental preservation infused with (primarily Cajun) history, such that the *relationship* between people and the swamp is preserved rather than a static sense of untouched wilderness. Writing stories of decline risks the continuation of environmental marginalisation, an outcome to which geographers and historians of wetlands in particular are resolutely averse (including Comeaux 1972, Vileisis 1997, Meindl 2000, Wilson 2006, Nelson 2009). If the goal of this text is to show that watery spaces of nature should be considered cultural as well, or to infuse the place image of water with that of 'habitable' land, the swamp stands as the metaphorical mud between concepts and is thus a vital space for inquiry.

LOSING THE SWAMP

Wetland historians and social theorists present loss within discussions of the broader progression of development on a regional or national scale but, in the end, the marsh is converted to a housing development (Meindl 2000), the swamp is drained and logged (DeBoer 2008) or the river basin is converted to monoculture farmland (Wilson 2006). These works have illustrated that unconstrained development and an unfortunate social stigma have combined to create an all-out assault on wetland spaces. This was particularly true in the US, where notions of 'sea to shining sea' effectively cast the in-between as land and spatialised swamps as 'fixer-uppers' in need of improvement. As Vileisis puts it, 'For centuries, the conviction that natural swamps and marshes were worthless and troublesome went hand-in-hand with people's actions and government policies' (1997, p. 7). However, since the growth of environmentalism (or as a result of it) overall values toward wetlands have shifted and allowed for protections in one manner or another throughout North America.

This broader story of destruction mediated by protection can also be applied to the Atchafalaya Basin, America's largest river basin swamp (a moniker created for its superlativity). Though in my research some developments in the swamp were considered by locals to be capable of 'killing' the ecosystem and its culture, others were seen as sustainable uses. I begin by reviewing the ecosystem-scale threats and continue by addressing the less-menacing uses of the swamp in the next section.

Especially following the 1927 Mississippi flood, immense levee projects were undertaken to rein in the distributary Atchafalaya River's natural outpouring into the swamps that surrounded it. The completion in 1963 of the Old River Control Structure at Morganza seemed to be the pinnacle moment where control was accomplished (Reuss 1998). No longer would major floods damage significant human settlements; however, this also meant the end of calmer yearly spring floods that renourished the outlying swamps. Reuss writes, 'Less water was available to rejuvenate the land behind the levees and more sediment was deposited further south where the levees stopped, turning some areas from swampland into farmland' (1998, p. 12). Since this Corps of Engineers project was completed, the Atchafalaya has essentially been controlled by computers upstream that ensure it captures thirty per cent of the Mississippi's flow and fifty per cent of its sediment.

The backfill in Atchafalaya hinterlands combined with over-sedimentation of the marshes nearer to the Gulf of Mexico has, by many accounts, incrementally caused the 'death' of this environment. Many of the informants from my 2010 research remembered times before the levees and a few were born in the Basin (as it is referred to locally) before it was 'controlled'. When asked, 'What is the biggest environmental threat?' several confirmed that siltation presented a threat of total loss. 'The Basin is silting in; that's a natural phenomenon', claimed one interviewee (Interview H 2010). This respondent's assertion that siltation is 'natural' cannot be refuted; as a river slows in its delta, silt is deposited that over time redirects the flow of waters. What made siltation a human-induced *problem* was the acceleration of this process. The Atchafalaya's siltation was sped up so that the Mississippi's could be slowed down, in order to facilitate transit.

Total physical loss of the ecosystem was also noted in the debate over sheet flows – shallow sheets of water that migrate down river, covering thousands of acres. When oil pipelines were dug through the flat swamp, heavy equipment was used to pile up mud (perhaps even as little as a few inches) for miles along the pipeline. Leaving what are known as 'spoil banks', this illegal practice impeded the flow of water, which became stagnant. Ecologically speaking, a healthy swamp should maintain a slow drifting water flow. Enforcement of the law or prosecution of offenders by the state was rare. Several guides and a representative from the crawfishermen's industry association noted that losing sheet flows was causing swamp loss and, more specifically, loss of prime crawfishing grounds. Ecosystem-level threats to the swamp were not always levied against only one culprit, but were more generalised. One guide put it bluntly, 'It's a tossup between the people living here and the ecosystem' (Interview D 2010).

USING THE SWAMP

When the above respondent presents the story as a competition between nature and culture, he seems to hint at the problem of imagining the space as the domain of one *or* the other. For centuries prior to the levees and environmentalism, people were more or less 'living off the [water]land', sometimes pillaging swamp commodities, other times sustainably harvesting foodstuffs, but maintaining a complex relationship where neither nature nor culture was the ultimate victor. In what follows, I suggest that, rather than only being an environmentally damaged space, the Basin is also a utilised space – a hybrid of nature and culture.

Some use-perspectives were rooted in historic and cultural/economic claims to the Atchafalaya – particularly as a space for crawfishing. Others feared that the more general conception of swamps as dirty, infested, scary or otherwise marginal places continued to motivate how they were used. This attitude was not espoused by the guides themselves, but was projected by them onto an anti-swamp 'other' who was constructed as the swamp's enemy. One indicative response from a guide was, 'People are treating it like a drainage ditch which it isn't and never has been' (Interview A 2010). This enemy presented a physical threat to the swamp as well as an ideological threat by imagining the Basin as a marginal, negatively connoted space, or by seeing the swamp only as space of capital investment and returns.

The Swamp: Waterland of the Semi-Wild

For many years, swamps were treated like a barrier to growth and capitalist development unless they were drained, logged and converted to farmland (Wilson 2006). The Atchafalaya still presented a barrier to development, but the attitude seemed to be that if a corporate entity invested enough time, labour and equipment in the Basin, it could reap financial rewards. In several interviews, respondents stated that money and corporate interests were more powerful than the law or the material trouble of getting to the resources. When speaking of levee-building and draining the swamp, one perturbed respondent said, 'The people that do that kind of thing usually have a lot of money. They can bring bulldozers and drag lines out there and change nature to their own wishes and everybody else can go to hell' (Interview I 2010). Others were more open to the idea of the swamp as a space of investment and return if development respected swamp ecology. 'Development and landowners have done their best to try to make money off of their land, which there is nothing wrong with, but when you destroy the environment to enable you to do that it is a problem' (Interview A 2010). Most respondents referred to 'development' as a well-funded, large-scale rearrangement of the ecology, particularly the flow of water. Development was generally seen as a corporate or big-money activity rather than development of crawfishing, tourism or other uses producing less environmental impact. The two most common sources of land-use development that the interviewees critiqued were farming and housing.

Soybeans and sugar cane were the crops most often seen as encroaching on the swamp. Though locals' involvements with some extractive industries (particularly oil) were seen as a point of cultural/historical pride, this was not the case for the cane fields, despite their longer history of cultivation in the Basin. A local who had seen the process of draining, logging and conversion to sugarcane on his generational family land (where he still lived) felt sorrow that he was too young at the time to prevent this from occurring. Another respondent countered development of soybeans against crawfishing, which he saw as more environmentally and socially sustainable.

> If you look at the numbers of productivity for crawfish per acre of the swamp it's amazing how many thousands of pounds can be produced. There is way more than if you drain the swamp and put a soybean field. So you can get more food per acre than any other terrestrial ecosystem and you have people that love it and really are willing to live in houseboats without hurting the environment taking advantage of the productivity of the swamp (Interview F 2010).

Agriculture was treated not only as an inappropriate re-assembling of the environment, but also as a source for pollutants that are detrimental to water quality and wildlife.

The second most commonly cited (ab)use of the swamp was draining for housing development. The pressure to suburbanise the edges of the swamp was most prevalent in the Baton Rouge area where the urban is separated from the swamp by only a few miles. Nowhere was this resentment more prevalent than at the now-defunct Alligator Bayou Swamp Tours south of the capital city. As revealed by a 'former employee, present volunteer' who still looked after on-site animal exhibits, the pressure to develop these areas for housing was magnified by the large migration of New Orleanians to Baton Rouge after Hurricane Katrina (Interview A 2010).

Resource extraction via oil drilling and logging were also problematic ways that locals saw the swamp being used. Oil exploration has been a staple on the Atchafalaya landscape since the 1940s, though, according to guides, only one well was still operational. Employment in oil has been maintained as a source of pride for some Cajuns – thus some respondents saw further exploitation of oil as an acceptable use of the swamp. One guide who had worked as a geologist stated 'We seismographed that whole Basin not long ago so there is oil in there but I don't see any reason why we shouldn't be drilling in the swamp' (Interview G 2010). However, most locals were opposed to oil drilling and the negative results of building pipelines through the swamp. These attitudes were probably exacerbated by the fact that, throughout my research, the *Deepwater Horizon* site was gushing thousands of barrels of oil a day into the Gulf of Mexico. The greatest impact of drilling noted by my interviewees was the preponderance of spoil banks left by the drilling companies. This problem was seen as especially detrimental to crawfishermen whose crop size has always been directly correlated to the amount of flowing oxygenated water. 'The frequent experience is like you go out and lift your traps and all the crawfish are dead because there is no oxygen in the water' (Interview I 2010). To fight this problem, the crawfishermen started an industry association (Louisiana Crawfish Producers Association) so they could pool their resources and voices and push litigation against the oil companies and the state for failing to enforce the law. Beyond this critical oil attitude, others were simply opposed to further drilling 'trashing it up with oil rigs' or felt that petroleum was doomed in the long run as an energy source (Interview E 2010).

Similar vitriol was directed at cypress logging, which was seen as unsustainable. Though cypress logging was outlawed in many areas of the Basin, small companies were able to set up operations that cut and chipped the trees for mulch sold at 'big-box' garden stores for as little as $2 a bag (Save Our Cypress Coalition 2011). Though illegal, these operations continued under the radar until 2008 when, motivated by indisputable aerial photographic evidence, the Corps of Engineers shut down the last mill (though the state did not prosecute) (Interview F 2010). One tour guide saw the elimination of logging on private lands as the primary reason that the government should buy up as much of the Basin as possible. In general, none of my respondents saw cypress logging as an acceptable use of the Atchafalaya, primarily because cypress trees take so long to regenerate.

The Basin was also constructed as a space of harvestable wildlife. The Atchafalaya National Wildlife Refuge covers over 15,000 acres in the heart of the Basin and allows public hunting, fishing and other outdoor recreation (Atchafalaya National Wildlife Refuge 2011). Many research participants saw this type of wildlife management as a sustainable use that also protected the cultural heritage of hunting and fishing. One guide felt that the management of and access to the Basin should be based on the historic rights of trappers: 'What built America? The frontier ... And the legislation was geared towards those people supporting that industry. So they built laws about trespassing in favour of the trappers' (Interview H 2011). Though trapping today is hardly an industry, to him, the usufruct rights of trappers were the roots of public land policy. Another respondent considered the Basin to be a place where wildlife management could co-exist with aesthetic beauty.

> It should be utilised; people ought to be able to commercial and sport fish it. And designate like they have today game preserves for hunting with a permit. I think it ought to be declared a national treasure or something in that category because it is Louisiana's Grand Canyon (Interview G 2010).

Many guides placed at least as much value on protecting the heritage of hunting and fishing as on the beauty and splendour of the environment. Though they did not specifically characterise it as such, the managed use of the swamp for renewable commodities was seen by guides as sustainable development.

All of my informants thought that the Basin should be utilised for the production of some commodities. No one felt that it should be walled off and left alone, which suggests that, at least locally, it is not treated as the domain of pure nature. This is not surprising though, since the swamp tour guides are themselves present on a daily basis. As I have shown, they were usually quite opinionated on the difference between sustainable use and outright exploitation.

FASHIONING THE SEMI-WILD

Though some locals' histories concur with the narrative of swamp decline, I have shown that, far from being dead, the swamp is utilised, disputed and protected. The social milieu of the Basin is as dynamic as its ecology and both seem to survive year after year despite outright threats or unsustainable uses. Particularly fascinating was the construction of the space as 'semi-wild' – a hybrid blend of such oppositional tropes as pristine wilderness and industrial oil drilling.

'There is something about being out in the swamp, especially when you're alone, that seems to suck the poisons of civilization out of you' (Interview I 2010). This palatable, healing power of place is appealing, yet, when theorised through the framework of literatures on space and natural purity, this perspective is itself poisoned. Pure nature may be fallacious but only in a theoretical sense, when informed by our understandings of history and ecology. In other words, a place could be experienced as 'untrammelled by man' – the definition of wilderness – if we had no knowledge or evidence that it had been spoiled (Wilderness Act, 1964). This, I believe, is the frame of reference that many visitors bring to the Basin, and for good reason. The swamp tour promotional pamphlets and websites show few pictures of the 'lake of stumps' landscapes or oil production platforms that dominate some areas of the Basin. Tourists often imagine the swamp as a wilderness but, for some swamp neophytes, the experience of a swamp tour can be disenchanting. One visitor from another state said, 'being that it said "Swamp Tours" I was expecting maybe a little bit more wilderness' (Interview B 2010). Guides negotiate this discrepancy by hybridising the purity of space so as to foster appreciation for what is 'left' and by enculturating the natural landscape with history. The narrative of civilisation in the swamp often becomes a story of local people living off the swamp and finally succumbing to the external pressures of resource development. This discourse is manifested through tours by establishing tropes of purity and ravaged loss, and finally by blending the two in the material context of the swamp. Though not explicitly mentioned on the boats, tourism then presents itself as sustainable use that maintains the hybrid nature/culture relationship.

Despite decades of logging, massive Corps of Engineers water control projects and petroleum extraction, the Basin is still constructed for tourism as a pure space of nature. One guide told me that in the places he goes, 'What you see is real' (Interview H 2010). To support the creation and management of the Atchafalaya National Heritage Area (ANHA), in 2008 the State of Louisiana conducted public information gathering sessions where they collected surveys from nearly 800 involved residents. Therein, commenters often used phrases that highlighted the purity of the space. Besides words like 'wild' and 'wilderness' they coded the Basin as 'pure', 'untouched' and a place where one could be 'close to nature' (ANHA comments 2009). One respondent's answers to the seven questions on the survey are telling and indicative of a common viewpoint:

> 1. Undeveloped wild area. 2. Undeveloped wild area. 3. Any human development. 4. See – undeveloped wild area 5. Undeveloped wild area. 6. Over development as tourist area. 7. Probably unrealistic but you asked (ANHA comments 2009).

Besides pure and pristine, many sources see the Basin as a place of beauty (which is hard to argue with). 'Breathtaking', 'beauty' and 'magnificent' were common impressions (ANHA comments 2009). These few references are aimed at the landscape in general, but many others tied beauty, wildness and purity to specific species such as cypress trees, eagles or alligators.

In contrast to tropes of purity, locals who participated in these information sessions also saw the Basin as a place where nature has been exploited and lost. Similar to my interviews with swamp tour guides, in the ANHA sessions people used phrases like 'destroying our wilderness' and 'return to the wild habitat it once was' (ANHA comments 2009). Just like the tour guides, other locals saw the Basin as semi-wild – a strange balance between destruction and pristine nature.

Semi-wildness is not a word that I came up with but one that emerged in several interviews independently and in the ANHA comments. Though several respondents referred to semi-wild specifically, others did so in a roundabout way by saying the Basin was a 'tiny microcosm of a wilderness'; 'in certain places it is a wilderness' or 'a wilderness in some ways' (Interview A 2010, Interview K 2010, Interview C 2010). In a theoretical sense the notion of semi-pure is impossible. However, neither wilderness nor purity *necessarily* imply a large scale of thousands of acres. Because of this, experiencing patchwork purity in the Basin seemed feasible. To designate which areas were pure and which were violated would be completely subjective and would ignore the preponderance of past activities that have altered the Basin on a grand scale, not to mention the fact that the Atchafalaya River's water level is constantly managed by the Old River Control. Yet the semi-wild still existed, despite it being theoretically impossible or objectively indefinable. Below I list a few interesting ways that locals gave credence to the semi-wild.

> 'I would like to see a few more interpretive sites and I wouldn't like to see it lose its "wildness"' (ANHA comments 2009).
>
> 'a relatively safe, wild-like experience' (ANHA comments 2009)
>
> 'Semi-wilderness; especially right here because the interstate is running through it' (Interview D 2010).
>
> 'Prior to the construction of the levees, you actually had communities living in the wilderness' (Interview E 2010).
>
> 'When you get out in the swamp it feels like you are in the middle of nowhere, but you know you are not' (Interview A 2010).

All of these cases reveal how people can create hybridities to deal with the inconsistencies of pure/ravaged spaces and, in the end, spaces of nature vs. culture.

In several instances, guides referenced the Interstate 10 bridge as a dominant emblem of the Basin's humanness. Several authors have commented on its significance as a 'conquering' of the swamp for transit (Bernard 2003, Reuss 1998). Paradoxically, another guide spoke of the opposite sense of space when he said, 'It is a wilderness area but that is only because you don't have roads and utilities and people living there' (Interview E 2010). Besides roads, people see its non-urban nature as indicative of its purity. One commenter brought out Cronon's (1996a) point by seeing value in 'the natural beauty that is not as evident in our towns and cities' (ANHA comments 2009). The Basin was also made wild by having a lack of people and containing 'dangerous animals' (ANHA comments 2009). It was constructed as less wild because of second growth forest, petroleum, canals, levies and tourists. These paradoxical negotiations of semi-wildness revealed disconnects between academic representations of space (via theory or tradition) and spatial practice.

CONCLUSION

In the Atchafalaya Basin, development has caused ecological degradation; yet the swamp is still alive and from many perspectives, well. Though a declensionist narrative of this space could seek out stories of unfettered development and wanton draining of the wetland, which are certainly accurate and fit into the mould for wetland environmental histories, I have shown how destruction and a lived notion of semi-wildness are juxtaposed by tour guides with lived histories of the swamp. We must recognise environments as lived and used both sustainably and unsustainably rather than only marginalising them as destroyed.

Throughout, I have used the word swamp rather than wetland. This was not intended as an appeal to specificity but as a critique on the present notion of wetland in common usage. Vileisis notes in her history of the nomenclature, 'In the 1950's ecologists coined the term *wetland* to replace the imprecise and value-laden *swamp*' (1997, p. 7). After sixty years of 'wetlands', this term has itself become value-laden, as it carries with it an imaginary of a threatened space, a space without culture, a space of objective ecology. For social researchers to acknowledge more than destruction, we must accept that environments are lived and appreciated, whatever their deviation from ecological constructions of balance or process.

The concept of semi-wild allows for destruction and appreciation to be considered in concert. It is drawn not from ecological histories, but from those who actually use and encounter the space. Applying this concept attempts to give voice to those environmental historical discourses that emerge *in situ*. Semi-wild recognises not that the environment was wild before humans came along and somehow tarnished it, but that nature has had culture all along. What has progressed is our *thinking* about places as wild and untouched. This can be an inroad to 'enculturating' all such spaces and in particular water-spaces where the division between aquatic nature and landed culture seems so distinct.

BIBLIOGRAPHY

Atchafalaya National Heritage Area Public Scoping Comments. 2009. Louisiana Department of Culture, Recreation and Tourism. http://www.atchafalaya.org/content/management-plan

Atchafalaya National Wildlife Refuge. US Fish and Wildlife Service. (2011) http://www.fws.gov/atchafalaya/ [accessed 20 April 2011].

Bernard, S. 2003. *The Cajuns: Americanization of a People*. Oxford, MS: University of Mississippi Press.

Comeaux, M. 1972. *Atchafalaya Swamp Life Settlement and Folk Occupations*. Baton Rouge: Louisiana State University School of Geoscience.

Cronon, W. 1992. 'A Place for Stories: Nature, History, and Narrative'. *The Journal of American History* 78: 1347–1376.

— 1996. 'The Trouble with Wilderness: Or, Getting Back to the Wrong Nature'. *Environmental History* 1: 7–28.

De Boer, T. 2008. *Nature, Business, and Community in North Carolina's Green Swamp*. Gainesville: University Press of Florida.

Ellis, J.C. 1996. 'On the Search for a Root Cause: Essentialist Tendencies in Environmental Discourse', in W. Cronon (ed.) *Uncommon Ground: Rethinking the Human Place in Nature*. New York: W.W. Norton & Co.

Interview A (2010) Swamp tour guide: 23 June.

Interview B (2010) Swamp tourist: 24 June.

Interview C (2010) Swamp tour guide: 7 July.

Interview D (2010) Swamp tour guide: 9 July.

Interview E (2010) Swamp tour guide: 13 July.

Interview F (2010) Swamp tour guide: 19 July.

Interview G (2010) Swamp tour guide: 19 July.

Interview H (2010) Swamp tour guide: 28 July.

Interview I (2010) Swamp tour guide, Louisiana Crawfish Producers Association Representative: 28 July.

Interview J (2010) Swamp tour guide, crawfisherman: 28 July.

Interview K (2010) Swamp tour guide: 3 August.

Interview L (2010) Swamp tour guide: 5 August.

Meindl, C. 2000. 'Past Perceptions of the Great American Wetland: Florida's Everglades during the Early Twentieth Century'. *Environmental History* 5: 378–395.

Nelson, M.K. 2009. *Trembling Earth: A Cultural History of the Okefenokee Swamp*. Athens, GA: University of Georgia Press.

Reuss, M. (1998) *Designing the Bayous, The Control of Water in the Atchafalaya Basin 1800–1995*. EP 870-1-53. Alexandria, VA: US Army Corps of Engineers.

Save our Cypress Coalition. 2011. http://healthygulf.org/our-work/save-our-cypress/save-our-cypress-home.

Vileisis, A. 1997. *Discovering the Unknown Landscape: a History of America's*

Wetlands. Island Press: Washington D.C.

Wilderness Act of 1964. Public Law 88-577 (16 U.S. C. 1131-1136) 88th Congress, Second Session 3 September 1964.

Wilson, A. 2006. *Shadow and Shelter the Swamp in Southern Culture*. Oxford, MS: University of Mississippi Press.

~ 2 ~

Filling Boston Commons: Law, Culture, and Ecology in a Seventeenth-Century Estuary

Christopher L. Pastore

Standing 138 feet above sea level, at the top of Beacon Hill, the Reverend William Blaxton could, in 1630, look down on the forests and rivers of mainland Massachusetts to the west and to what later settlers would call Boston Bay to the east (Rutman 1965, p. 28; Amory 1880, pp. 6–7). To the south, he could see the thin thread of an isthmus, flanked by tidal salt marsh and mudflats, connecting the largely treeless, mile-wide Shawmut peninsula to the mainland (Vileisis 1997, p. 30; Whitehill 1968, pp. 3–5). The neck, 'forty perches large', observed William Harris later in the century, was so narrow that it was often inundated during storms 'over a low-salt marsh & wch is sometimes over flown by ye Sea' (1902, p. 144). On the neck's western side was the Back Bay, dominated by Round Marsh. And farther north was Blaxton's Cove, which was also ringed with marsh grasses. On the northern tip of the peninsula was the Mill Cove, a shallow bay at the mouth of which a sandy, grass-covered bar had formed as seabed sand was piled there by the tide and silt and seeds from the peninsula travelled down its spring-fed streams. On Shawmut's north-east corner was the Great Cove, also known as the Town Cove, on the western corner of which was 'the creek', a tidal marsh channel running westward to the Mill Cove. Finally, on the peninsula's south-east corner, reaching to the east side of Shawmut neck, was South Cove, which was similarly dominated by the grassy mats and sprawling flats that shaped the greater Charles River estuary (Rutman 1965, pp. 32–40; Whitehill 1968).

When Jonathan Winthrop with roughly 700 settlers arrived later that year, Blaxton invited them to share his watery world. They named their town Boston after the town in Lincolnshire, England whose landscape was similarly dominated by sweeping fens and from which several members of the Winthrop party had hailed (Dudley 1630, pp. 309, 313). Like its

English namesake, Boston depended on its tidal landscape, for the nascent town quickly became the hub of an intricate system of exchanges between land and sea. Its settlers fished for cod, herring, sturgeon and salmon and trapped beaver, martin and otters (Wood [1634] 1865, pp. 27–9, 36–7). They felled forests and planted crops. The high production of colonial materials in the first half of the seventeenth century increased overseas investment in the colonies, which induced more Europeans to settle there (Bailyn 1955, p. 15). In the marshes and mudflats, newcomers took their first steps on New England soil, wading through waist-high grass, their feet bogging in black mud. To facilitate this intertidal traverse, they built paths and wharves, changing nature to meet their needs.[1]

Just as marshes and mudflats served as Boston's commercial interface, they likewise formed a biological one, an ecological margin marshalled by wind, tide and the effects of storms and scouring ice. In this intertidal transition zone, or ecotone, freshwater and saltwater plants wrestled for dominance. Terrestrial animals preyed on marine animals and vice versa. For both land and sea (and the organisms that lived in these environments), the salt marsh and tidal flats were coterminous and, as a result, they were biologically rich. Spawning alewife, shad and salmon clogged tidal rivers. Shellfish beds fed settlers and Native Americans alike. And marsh hay fed their animals.

The ecological intricacies of the estuary also shaped the legal landscape of Boston's inhabitants, particularly with respect to property ownership. On one side of the estuary rested deeded farmlands and on the other, a legally unruly ocean. And somewhere in between, in the muddy, brackish pools left with each outgoing tide, swirled complex English assumptions about land and sea, private ownership and common usage. To obviate confusion, a 1641 Massachusetts law drew the line between land and sea, extending ownership rights to the extreme low-tide mark (Whitmore 1887, p. 91). Intended to promote wharf construction by deeding tidal flats to builders, the ordinance nevertheless upheld common right of way of that same land for fishing, fowling and navigation. What does this law, one that sought to mesh private land ownership with common ocean ownership, say about settler conceptions of the estuarine environment? And what was so sacred

1 From the early seventeenth century to the present day, Boston has moulded its tidal lands to meet human demands, filling vast areas for seaports and airports, wharves and wealthy neighborhoods. See Seasholes 2003.

about that low tide mark – where land ended and the ocean began – that legislators decided not to extend ownership farther seaward?

The concept of deeded land ownership that concomitantly granted usufruct rights had, for some English settlers, been unsettling. This shared-use model sometimes became a point of contention when negotiating land deals with Native Americans, who, in ceding lands, often insisted on maintaining hunting, gathering and fishing rights (Cronon 1983, p. 74; Leavenworth 1999, pp. 275–300). Although some regions in England had maintained systems of property ownership that also allowed common usage, the developing Anglo-American land-tenure system encouraged private proprietorship (Allen 1981; Stilgoe 1982, p. 54; Fischer 1989, pp. 31–54). This settler conception of ownership typically granted exclusive right of use and stringent rules against trespassing (Cronon 1983, p. 73). The 1641 law, however, codified a private–public sharing system. But as Boston's trade increased and available waterfront land diminished, merchants sought to consolidate their landholdings, which often required them to block their tidal lands from the sea's innate ability to make them common (Bailyn 1955, p. 102). To stem the tide they dyked marshes and filled between wharves. They physically built a more definitive line between land and sea, which likewise sharpened the conceptual line between private ownership and common usage (Cronon 1983, p. 74).[2]

It is my contention that it was in the place where two biomes met (to form a third), where freshwater and saltwater mixed and where plants and animals were forced to adapt to a landscape in perennial flux, that humans, too, developed adaptations to the salt marsh's complex ecology in the form of cultural assumptions and laws that braided their conceptions of land and sea. Ultimately, this chapter argues that the changing ecological relationship between Boston's early English settlers and their estuary was instrumental in shaping Anglo-American conceptions of property rights that favoured private ownership over shared-use models.

2 Cronon (1983) explains that the earliest land contracts 'had a superficial resemblance to Indian usufruct rights, since they seemed to define land in terms of how it was to be used'. But he points out that 'later deeds described land in terms of lots held by adjacent owners, and marked territories using the surveyor's abstractions of points of the compass and metes and bounds'.

ENGLISH OWNERSHIP ON LAND AND SEA

When early seventeenth-century English settlers populated the various towns of Massachusetts Bay, they brought with them landholding traditions that had prevailed in their respective home regions. In his study of the towns along Boston's North Shore, David Grayson Allen showed that such factors as religious custom and social structure, terrain and technique shaped community modes of land usage. The community of Rowley, Massachusetts, for instance, imported an open-field system from Rowley, Yorkshire, where farmers owned non-contiguous patches of land that fell under common regulation. In Hingham, Massachusetts, as in Hingham, Norfolk, dairy farming was supplemented by growing small amounts of grain. Both Hinghams combined an open-field system with one that included enclosed property ownership. The settlers of Newbury hailed from the Wiltshire-Hampshire region, which had a tradition of central villages surrounded by common fields and pastures. Cultural and religious differences in the Wiltshire-Hampshire region led to wide variation in land usage, so that settlers of Newbury developed an eclectic mix of enclosure and commonage. These differences caused numerous land disputes in Newbury, which pushed the town toward private proprietorship. Similarly, for the settlers of Ipswich and Watertown, who had largely come from East Anglia, land enclosure was the custom (Allen 1981, pp. 22, 57–64, 115–16, 160).

Over time, however, these land practices changed. By the final decades of the seventeenth century, the various land-management systems grew more alike as decision-making power gravitated away from individual towns and toward the Massachusetts Bay General Court (Allen 1981, pp. 223–4). As a more unified conception of land ownership developed under the General Court authority, tidal lands, in particular, were subjected to sharp scrutiny. While grappling with the cultural, ecological and legal complexities of New England's terrestrial farmlands, early settlers transplanted analogous assumptions to their salt marshes and mudflats. These assumptions, in turn, played a large part in shaping the laws that governed the ways the edge of the sea was partitioned.

THE LAW OF THE SEA

Such questions concerning ocean ownership were not limited to the local level. As European nations expanded their land holdings during the sixteenth century, many claimed dominion over the surrounding seas. To refute Spanish claims on the Pacific Ocean and Gulf of Mexico, and those of Portugal, which claimed the Atlantic south of Morocco and the Indian Ocean, in 1608 the Dutch jurist Hugo Grotius published his *Mare Liberum*, or *Freedom of the Seas*, arguing that, since antiquity, the ocean had been considered a commons and that it should remain so (Grotius [1608] 1916, p. vi). Voicing support for Dutch commercial shipping access, Grotius contended that no governmental law could grant ownership of the sea. The commons component of the sea, he contended, was rooted in nature. To claim ownership of the ocean was an affront to God.

For Grotius, the freedom of the sea applied to coasts as well. '[T]he same right which applies to the sea', he argued, 'applies also to the things which the sea has ... made its own, such ... as the sands of the sea, of which the portion adjoining the land is called the coast or shore' (Grotius [1608] 1916, p. 28). Any portion of land inundated by the sea, such as beaches or tidal marshes, should be held in common. But, he explained, 'If any part of these things is by nature susceptible of occupation, it may become the property of the one who occupies it only so far as such occupation does not affect its common use' (p. 30). If one built, say, a fence or wharf, asserting inhabitance of tidal land, it could be claimed as private property as long as it was still made available for public use. But in 1635 John Selden, an Englishman, published a rejoinder titled *Mare Clausum*. He avowed, 'That the Sea, by the Law of Nature or Nations, is not common to allmen [*sic*], but capable of private Dominion or proprietie as well as the Land'. Selden's ocean was *clausum*, or capable of enclosure. And who owned it? '[T]he King of *Great Britain* is Lord of the Sea flowing about, as an inseparable and perpetual Appendant of the British Empire' (Selden 1652, pp. e2-f).

Although both Grotius and Selden were concerned with ocean ownership in terms of competing nations as opposed to individual property claims, the ideas on which they predicated their arguments trickled into the legal discourse that shaped waterfront property regulations in Massachusetts Bay. Were the ocean and its tidal arms free or could they be owned? And if they were capable of private possession, were there limits to the title? Although most early seventeenth-century English subjects would probably agree with Selden that the North Atlantic between England and New England was the King's property, when considering ownership at the local level, their customs and later laws ultimately reflected the views of Grotius. Tidal property was common but, when bounded by a surveyor's stick, it could be owned as well.

There is some evidence that suggests early settlers were confronting the legal complexities of littoral ownership before specific laws had been adopted. In describing 'a spacious Marsh' on Nahant Neck, William Wood noted in 1634 that a 'few posts and rayles from the low water-markes to the shore' had been erected to protect cattle, 'weather-goats, and Swine' from 'Woolves' (Wood [1634] 1865, p. 46). That fencing had been erected below the tide suggests these marshland farmers had invested some effort to bound out and claim proprietorship of the shore. Wood noted that the property was given to the 'plantation of *Saugus,* so that no other can appropriate it to himselfe', which suggests there had been or the Saugus farmers anticipated conflicting claims on their intertidal land (p. 46).

Reflecting this low-water local custom, and designed in part to maintain amity alongshore, the Colonial Ordinance of 1641 codified the colony's policy on waterfront property. The law stated, 'That in all Creeks, Coves and other places about and upon Salt water, where the Sea Ebbs and Flows, the Proprietor, or the Land adjoyning, shall have Propriety to the Low-water mark, where the Sea doth not Ebbe above a *hundred Rods,* and not more wheresoever it Ebbs further' (Whitmore 1887, p. 91). This new law marked the lowest tide – the line closest to the sea – as the outward boundary of private ownership. Yet the law still granted public access to this intertidal land. A farmer could wander onto his neighbour's tidal marshland and mudflats to fish, hunt and navigate safely to sea, but the neighbour still held legal title to the land. By deeding tidal flats to builders, the law also provided an incentive for seaward expansion. A builder would be more likely to erect a wharf and maintain it if he held legal title to the land on which it was built. But as property values increased and paling upland terrain became common practice, settlers increasingly made changes to their littoral lands that removed the commons component from them.

By the 1640s, Boston's sea trade had increased dramatically, which spurred waterfront development. Commercial expansion combined with the Colonial Ordinance of 1641 encouraging wharf construction, which sent a ripple of activity through Boston Harbor. As early as 1633 the town landing at the head of the Town Cove had been filled to facilitate boat landing. In 1638 fourteen investors built a wharf, crane and warehouse, which became the town dock in Benall's Cove. By 1648, they had filled areas on either side of the cove to further facilitate landing. By 1676, this cove was entirely enclosed, barring a swing bridge at its mouth, which connected the north and south sides of the harbour's western shore (Seasholes 2003, pp. 3, 22–5). The General Court approved numerous smaller wharfing projects as well. On 26 April 1647, both Benjamin Ward and John Anderson, the General Court decided, 'hath libertye to wharf afor his proprietye' (Whitmore and Appleton 1881, pp. 90–1). On 1 March 1650, eight waterfront owners were granted permission to 'wharfe or peare before his proprietie to low water marke' (Whitmore and Appleton 1881, p. 102). Wharf building reshaped Boston's littoral, for, as more and more wharves were built and longer wharves extended farther over the mudflats, in many cases, the land between them was filled in a process called 'wharfing out' (Seasholes 2003, p. 3).

Intertidal filling had become compulsory in a town whose available land was rapidly dwindling. As Bernard Bailyn has shown, 'English merchants bought and sold, bequeathed and inherited, mortgaged and released land in a bewildering maze of transactions' (Bailyn 1955, p. 103). In Boston, both a hub of local and international trade, waterfront property was particularly valuable. As available parcels grew fewer and more expensive, tidal landowners consolidated their holdings. By building wharves and filling between them, they claimed the commons as their own, creating land that could be passed to one's heirs in perpetuity. 'And whoever shall have constructed a building under the aforesaid circumstances will become the owner of the ground upon which said building is', wrote Grotius of the intertidal zone, 'because this ground is neither the property of any one else, nor is it necessary to common use'. Even the great champion of the freedom of the seas conceded that building upon tidal land would solidify private proprietorship. Construction would strip the commons from tidal lands, making moot the finer points of the Colonial Ordinance of 1641 and converting waterfront property to its owners free and clear. But Grotius also exposed the threat of inaction. '[H]is ownership lasts no longer than his occupation lasts, inasmuch as the sea seems by nature to resist ownership. For just as a wild animal, if it shall have escaped and thus recovered its natural liberty, is no longer the property of its captor, so also the sea may recover its possession of the shore' (Grotius [1608] 1916, p. 30). Privately deeded tidal land would remain so, but without having fortified one's land from the commons-creating effects of the ocean, ownership became conceptually and legally tenuous.

In a competitive commercial environment in which those with capital and political influence manoeuvred aggressively for waterfront property, inaction by Boston owners could weaken their claims to their land. On 1 April 1634, the General Court ruled 'that if any man hath any … land graunted him & doeth not builde upon it or imp[rove it] within Three yeares, it shalbe ffree [*sic*] for the Court to disp[ose] of it to whom they please' (Noble 1904, p. 43). In 1657 the General Court ordered that tidal landowners at Mount Wollaston and Muddy River who had not yet bounded their land were required to register with town officials or risk forfeiting ownership (Whitmore and Appleton 1881, p. 135). As Grotius had argued and the General Court mandated, property owners had either to use it or lose it. A failure to improve tidal lands could in some cases abrogate claims to private ownership. This likely provided an incentive for property owners to develop waterfront lands in ways that removed the commons component from them. The process of wharfing out, like paling uplands, marked a shift away from an earlier mixed private–public ownership system to one in which private proprietorship was the norm.

Failure to develop one's tidal lands could also result in their seizure by neighbours. Waterfront property owners who failed to wharf out at the same pace as those next door could lose their harbour frontage as the surrounding shoreline pushed farther away from them. In some cases where owners extended property lines over tidal mudflats, particularly along stretches of undulating shore, conflicting claims resulted. Although waterfront property owners could build 'no Broader [across the tidal flats] then his ground is at hie water Marke', there was always the potential that one set of property lines extending seaward could cross those of another nearby (Whitmore and Appleton 1881, p. 102). By the early nineteenth century Massachusetts courts were forced to re-examine the 1641 Colonial Ordinance for this very reason (Bowditch 1842, pp. 1–40).

In response, the processes of filling between wharves continued. Instead of gradually sloping tidal flats, which allowed the sea to creep for, in some cases, hundreds of yards, landowners piled rocks and soil between their piers, extending their property seaward and reducing the reach of the tide. They dyked marshes and built walls, extending exclusive private ownership seaward. By building a more definitive line between land and sea, they likewise sharpened the conceptual line between private ownership and common usage. Removing the intertidal ecotone, the transitional space

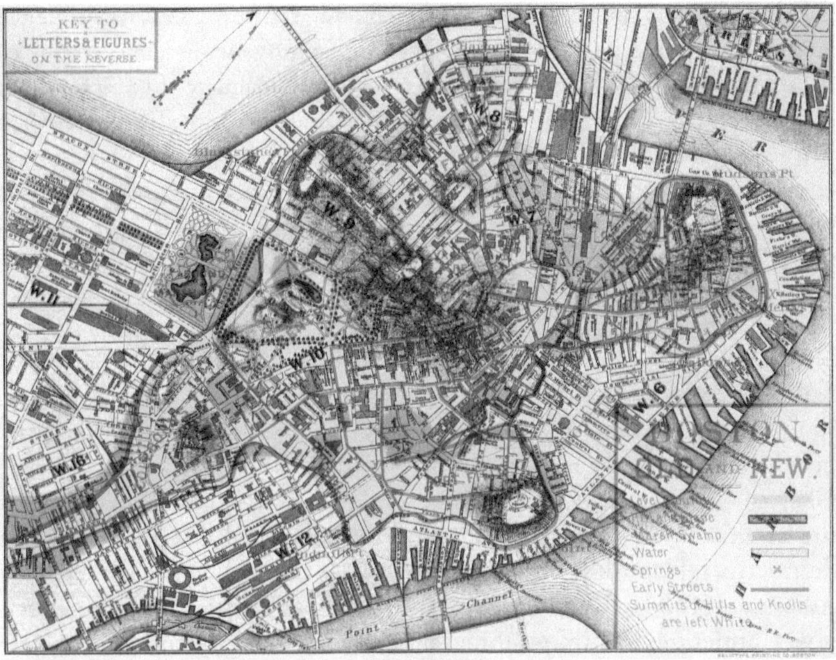

Figure 1. *Boston Old and New*

A 1641 Massachusetts law that extended property ownership to the low-tide mark while concomitantly maintaining public access for fishing, fowling and right of way encouraged the people of Boston to fill their intertidal zone, thereby constructing a more definitive line between land and sea. Between the seventeenth (see the inset coastal outline) and nineteenth centuries they systematically removed the commons-creating forces of the ocean. This transformation of the ocean's edge played a role in forging Anglo-American notions of property rights that favoured private ownership over shared-use models. Justin Winsor, *Boston Old and New* (Boston: Heliotype Printing Co., 1880). Map image courtesy of the Norman B. Leventhal Map Center at the Boston Public Library.

between land and sea, changed both the ecology of the estuary and the way people interacted with it. Settlers who had once shared a flexible understanding of land tenure, who were open to systems that shared common lands or, as in the tidal marshes and mudflats, supported a private–public mix, forged a more exclusive understanding of proprietorship when they set out to reclaim the tidal landscape from the sea. As the Massachusetts Bay

General Court built a more unified legal code that extended its influence across the colony, it likewise shaped more homogeneous conceptions of property ownership. Paradoxically, the Colonial Ordinance of 1641, which sought to meld colonial conceptions of land and sea by introducing a hybrid private–public ownership structure, ultimately prompted Massachusetts coastal settlers to physically and conceptually construct more permanent boundaries between the sea and shore, which in turn hardened their once more flexible conception of property ownership.

CONCLUSION

Much has been written about how early settlers sought to tame their environment. Some scholars have attributed this assault on nature to religious beliefs and others to capitalism (White, Jr. [1967] 1968, Oelschlaeger 1991, Glacken 1967). Seventeenth-century thinker Thomas Tryon believed it was the latter, contending that market forces imbued Englishmen with a more destructive relationship with their environment compared to that of Native Americans (1684, pp. 151–3; Cronon 1983, p. 170). Karl Marx contended it was a mixture of both, positing that the Christian conception of private property and a specie economy had led to unparalleled environmental exploitation, or the 'civilizing influence of capital' ([1939] 1993, p. 409). This essay contends that settlers were driven to modify the shoreline not simply out of divine duty or for want of money. Rather, their relationship with the intertidal zone was shaped by a more complex combination of ecological factors and cultural assumptions about land, sea and the muddy spaces in between.

As settlers grappled with the conceptual complexities of coasts, they created specialised laws. But reconciling the duality of tidal space proved difficult, for early colonial understanding of land ownership varied from town to town and from person to person, depending on the region of England from which he had emigrated. Early colonial conceptions of the sea were likewise in flux. The General Court's 1641 Colonial Ordinance, which maintained public usufruct rights for fishing, fowling and navigation while nevertheless granting private ownership of tidal lands to the lowest tidemark, reflected this broader cultural and legal ambivalence toward the status of this liminal intertidal space. The Ordinance granted one the ability to own the land on which a wharf was built while concomitantly extending Boston's commons into that same area. In an environment where waterfront land was quickly gaining value, this shared-use model was an affront to a growing settler sensibility that increasingly prized exclusive private ownership. By the mid 1640s, the English preference for outright ownership, in concert with the law, induced rapid ecological change. As colonists sought to purge their property of its common component – the commons influence of the sea – they built dykes and dams and filled between wharves. Ultimately, this physical and conceptual construction of a more succinct line between land and sea, one that thwarted the commons-creating effects of tidal flushing, played an important role in shaping a more rigid conception of property ownership in colonial America.

BIBLIOGRAPHY

Allen, D.G. 1981. *In English Ways: The Movement of Societies and the Transferal of English Local Law and Custom to Massachusetts Bay in the Seventeenth Century.* Chapel Hill: University of North Carolina Press.

Amory, T.C. 1880. 'William Blaxton', in *The Bostonian Society Publications* 1, no. 1. Boston: Old State House. pp. 6–7.

Bailyn, B. 1955. *The New England Merchants in the Seventeenth Century.* Cambridge, MA: Harvard University Press.

Bowditch, N.I. 1832. *Wharf Property: The Law of Flats.* Boston: John Wilson & Son.

Cronon, W. 1983. *Changes in the Land: Indians, Colonists, and the Ecology of New England.* New York: Hill & Wang.

'[Thomas] Dudley's Letter to the Countess of Lincoln', 12 March 1630. In Alexander Young (ed.) 1846. *Chronicles of the First Planters of the Colony of Massachusetts Bay, from 1623–1636.* Boston: Charles Little and James Brown.

Fischer, D.H. 1989. *Albion's Seed: Four British Folkways in America.* New York: Oxford University Press.

Glacken, C.J. 1967. *Traces on the Rhodian Shore: Nature and Culture in Western Thought From Ancient Times to the End of the Eighteenth Century.* Berkeley: University of California Press.

Grotius, H. 1916. *Mare Liberum* [1608], in J.B. Scott (ed.) *The Freedom of the Seas or The Right Which Belongs to the Dutch to Take Part in the East India Trade* trans. R. Van Deman Magoffin. New York: Oxford University Press.

'Harris's Account of New England, 29 April 1675'. 1902. In *Collections of the Rhode Island Historical Society,* vol. 10. Providence: Rhode Island Historical Society.

Leavenworth, P.S. 1999. '"The Best Title That Indians Can Claime": Native Agency and Consent in the Transferal of Penacook-Pawtucket Land in the Seventeenth Century'. *The New England Quarterly* **72**: 275–300.

Marx, K. [1939] 1993. *Grundrisse: Foundations of the Critique of Political Economy*, trans. M. Nicolaus. New York: Penguin.

Noble, J. (ed.) 1904. *Records of the Court of Assistants of the Colony of the Massachusetts Bay 1630–1692*, vol. 2. Boston: County of Suffolk.

Oelschlaeger, M. 1991. *The Idea of Wilderness: From Prehistory to the Age of Ecology.* New Haven: Yale University Press.

Rutman, D.B. 1965. *Winthrop's Boston: A Portrait of a Puritan Town, 1630–1649.* Chapel Hill: University of North Carolina Press.

Seasholes, N. 2003. *Gaining Ground: A History of Landmaking in Boston.* Boston: MIT Press.

Selden, J. 1652. *Of the Dominion, or Ownership, of the Sea.* London: William Du-Gard.

Stilgoe, J.R. 1982. *Common Landscape of America: 1580–1845.* New Haven: Yale University Press.

Tryon, T. 1684. 'Philotheos Physiologus', *The Country-Man's Companion.* London: A. Sowle.

Vileisis, A. 1997. *Discovering the Unknown Landscape: A History of America's Wetlands*, Washington, DC: Island Press.

White, L., Jr. 1967. 'The Historical Roots of our Ecologic Crisis', *Science* **155**, reprinted (1968) as Chapter 5 of *Machina ex Deo: Essays in the Dynamism of Western Culture.* Cambridge: MIT Press.

Whitehill, W.M. 1968 [1958]. *Boston: A Topographical History.* Cambridge, MA: Belknap Press of Harvard University Press.

Whitmore, W.H. (ed.) 1887. *The Colonial Laws of Massachusetts.* Boston: Rockwell and Churchill.

Whitmore, W.H. and W.S. Appleton (eds). 1881. *Second Report of the Record Commissioners of the City of Boston; Containing the Boston Records, 1634–1660 and the Book of Possessions.* Boston: Rockwell and Churchill.

Wood, W. 1865. *New-England's Prospect* [1634]. Boston: Prince Society.

~3~

Escaping the Maritime Revival Viewpoint

Glenn M. Grasso

In Gloucester, Massachusetts, it is a rare fisherman's pickup truck that does not sport a bumper sticker reading 'National Marine Fisheries Service: Destroying Fishermen and Their Communities since 1976'. These stickers, like dandelions after rain, sprout up in coastal Massachusetts, New Hampshire and Maine. Often, they are affixed to vehicles driven by men who had not yet been born in 1976 – or 1986, for that matter. Despite cod fishing's 1980s crash, Reagan-era over-capitalisation and debt, overwhelming scientific evidence that stocks are dwindling worldwide and witnessing their fathers' struggles firsthand, these fishermen stubbornly persist in an industry that would not elicit one iota of sympathy from the public, politicians or the fishermen themselves, were it not seagoing.

Contemporary fishermen benefit from a strong strain of romanticisation about their imagined historic lifestyle.[1] This phenomenon I have termed the Maritime Revival, and its cultural influence is far-reaching when analysing most marine topics (Grasso 2009). The Maritime Revival affects almost every seagoing trade, although some are privileged over others. A perceived decline in post-Civil War American sailing vessels resulted in a small subset of the maritime experience being lauded as heroic and imbued with essential Americanism. In actual fact, American tonnage rose during this era. This period also witnessed a great influx of southern and eastern European immigrants, growing urbanisation and massive industrialisation on scales previously unseen. Old-stock Americans responded by locating core values in pieces of history such as the Revolutionary generation, early colonists, architecture and seafaring – particularly square-rigged sailing. Revivalism helped old-stock Americans buttress their values and emphasise their own good tastes while offering educational examples to the immi-

1 Consider the influence of Sebastian Junger's *The Perfect Storm* and the Discovery Network's popular television series *Deadliest Catch.*

grants. Certain activities were elevated, their values set as benchmarks; and these corrected, sanitised and ultimately incomplete pieces of the past were distributed throughout American culture via art, advertising, literature, museums and experiential learning.[2] From about 1870 to 1890, Maritime Revivalism concentrated on square-rigged sailing but, after 1890, its focus expanded to include fishing and whaling.

The process designated some working-class experiences as 'authentic' lifestyles to be elevated, co-opted and explored by middle and upper classes, but the experiences-as-lived were sanitised.[3] By creating a carefully-controlled image of the sea and seafaring and encouraging its cultural diffusion, the Maritime Revival has, among other things, helped Americans discover a marine aesthetic, seaside holidays and the desirability of eating fresh seafood. This cultural construction has had significant consequences for the marine environment.

FISHING'S IMAGE

Commercial fishing is sometimes colloquially called 'harvesting the sea', invoking quiet images of a preindustrial, pastoral labour ripe with the Protestant work ethic. Harvesting is a misnomer – the image false – because harvesting implies cultivating. Fishing should more accurately be termed mining; indeed the closest analogy for an industry extracting what is already there. Mining conjures images of tunnels devoid of sunlight, nasty work, low wages and constant fears of collapse. Once a vein of ore is depleted, workers understand that mines will close, jobs will leave and local economic ruin is not only possible, but likely.

Fishing stands in stark contrast. Despite the reality of collapse, fishermen and sympathetic politicians support increased efforts (Urbon 2011, Kerry 2011). Each closing, quota or regulation prompts the industry to dig in, determined to sustain itself and its communities by employing a shared common heritage as rationale. Fishing regulations are 'most of all, taking our children's future away', proclaimed Fran Beaulieu of Rockport, Massachusetts in 2010 (Beaulieu 2010). 'This is not the America our fathers, grandfathers, aunts and uncles fought and died for.' Because their forefathers

2 See Rhoads 1977, Axelrod 1985, Morison 1921, Denenberg 2003, Lynes [1954] 1983, Gombrich 2002 and Gigante 2005.

3 Lears 1981 provides the best assessment of searching for authenticity.

fished, they conclude it is their birthright, too. More likely, their ancestors engaged in combinations of fishing, day labouring and farming – none privileged above another – to make ends meet.[4]

Beaulieu's opinion is hardly unique. Many fishermen insist there are plenty of fish and regulators are ignoring the facts. To be fair, there have been problems with bad data and computer models. Nevertheless, increased fishing effort – bigger boats, more gear and onboard refrigeration – is the principal reason landings continue to increase. Compared to efforts, catches (per unit of effort – CPUE) are declining, and have been for decades, even centuries. What fisheries scientists term the 'shifting baseline syndrome', a phenomenon where each generation perceives abundance despite real intergenerational change, compounds the problem (Pauly 1995, p. 430). Michael Conathan, Ocean Policy Director at *American Progress*, claims the ocean has an 'image problem' because everything looks in order to casual observers or beachgoers. Since we cannot see beneath the surface or at a microscopic level, we fail to consider habitat degradation, pollution or ocean chemistry. 'It's hard to ask people to develop a different perception than what they see with their own eyes', Conathan concludes. 'If a picture is worth a thousand words, an experience must be worth a billion'.[5] With typical human memory only going back a generation or two, failing to grasp the severity of the crisis and cherry-picking data are understandable, especially for those whose livelihood depends on fables of abundance.[6]

Nevertheless, no industry guarantees jobs in a business that is economically unviable. To paraphrase Ross Perot in 1992, if you make buggy whips, you better make the best buggy whips in the world. No one and no government promises anyone a job simply because one's father worked it. Whether California sardines, Pacific salmon, Atlantic halibut or Gulf of Maine cod, unregulated fishing's effects are well documented. Left to its own devices, the fishing industry would suffer the same fate as a depleted mine: economic ruin; and the now-former fishermen would need to find employment elsewhere.[7] Despite fishermen's protests, regulation with the goal of maintaining fish stocks offers an alternative not available to miners.

4 See Vickers 1994 and Vickers and Walsh 2005.

5 Conathan 2011. See Roberts 2007, Jackson, et al. 2001, Jackson 2001 and Berrill 1997.

6 See McClenachan 2009. With thanks to Jackson Lears for the term 'fables of abundance'.

7 Marine environmental history is rife with examples of commercial extinctions resulting from unregulated fishing. See McEvoy 1986, Taylor 1999, Grasso 2008 and Rosenberg et al. 2005.

Behind this bleak picture a central question remains: what makes commercial fishing different from other extractive industries and cultivates public and political sympathies? Is there something inherently different about fishing, fishermen and their communities compared to buggy whip manufacturers and theirs? Or, has a romantic image of commercial fishing been constructed over the past 150 years; one that stresses a heroic narrative worth remembering while ignoring changing technologies and environmental degradation? Through the lens of the Maritime Revival, I argue that this is precisely the case, and not only for the fishing industry.

Always a dangerous occupation, fishing communities used the danger to heroicise industry participants. Gloucester's Procter brothers published nineteenth-century memorials to honour the dead and fishermen were seen as heroes in literature, art and advertising (Babson [1860] 1972, p. 597; Procter 1873, p. 8; Procter Brothers 1882, p. 85). Rudyard Kipling's *Captain's Courageous* (1897) identified Gloucester fishermen's values as fundamental and relocated them to transform spoiled brat Harvey Cheyne. Based on Augustus Buhler's 1901 painting *Man at the Wheel,* Gorton's Fish Company adopted the heroic fisherman in a yellow Sou'wester as its logo for national marketing in 1904. The Fishermen's Memorial statue by Leonard Craske further reified the industry's heroism, honouring the '10,000 Gloucester fishermen' lost by 1925, and the community continues to add names.[8] As with memorials to other tragedies, there is comfort in remembrance. Against this backdrop of exceptionalism, challenging someone's commercial fishing birthright becomes akin to defacing a war memorial.[9]

The line from memorials to art suggests that fishing's public image bore little resemblance to Gloucester's 1920s fleet. The remaining sail-powered schooners fished just enough to remain eligible for organised races 'crewed' mostly by captains (Santos 2002). Fishing was mechanised and even the old schooners shortened their rigs and added engines. Despite modern methods, fishing's quaint preindustrial image held, encouraging tourism in New England's ports. This culturally-constructed image powerfully affected how politicians and the public viewed commercial fishing. Emotion superseded

---

8 Rocky Neck Historic Art Trail website and North Shore Community College website.

9 Gabbatt 2010. Charlie Gilmour, son of Pink Floyd guitarist David Gilmour, earned shame and arrest when he swung on a flag at London's Cenotaph war memorial during a December 2010 tuition fees protest. He quickly issued an embarrassed apology.

modern fishing's reality, and the industry retained a positive reputation, greatly influencing public policy decisions – and ecosystems.

WHALING'S IMAGE

Whaling's historic image falls squarely within the Maritime Revival viewpoint, but with an effect opposite to that of commercial fishing. Whaling gained heroic status in memory as it declined but, in its heyday, long voyages, low pay and gruelling work rarely made whaling a mariner's first choice. Yet, by the 1920s, a whaleship had been exhibited at the 1893 World's Columbian Exposition; statuary appeared in New Bedford; Clifford Ashley's paintings and book, *The Yankee Whaler*, had been published; Whaling Enshrined preserved the whaleship *Charles W. Morgan* in Round Hill, Massachusetts; and forgotten author Herman Melville enjoyed something of a renaissance (Cooney 1950, pp. 19–20; Lindgren, 1999; Ashley 1926; Leavitt 1973; Weaver 1921). Melville's case is doubly-noteworthy: not only did *Moby-Dick* bring heroic whaling into modern America, but Melville's Dutch heritage represented old-stock American artistic expression in a culture eager to cast off Victorian trappings in favour of 'real' American blood.

Sublime in their enormity, whales practically define the term 'charismatic megafauna' and, while romanticising commercial fishing resulted in its persistence, revivalism regarding whaling led its modern image down a different path. As Greenpeace publicised bloody photographs of mechanised whaling and the International Whaling Commission's 'blue whale unit' decimated every species it was designed to protect, public pressure mounted to end commercial whaling (Ellis 1991, pp. 365, 392, 426–8). However, the heroicised values of nineteenth-century whaling remained fixed in public memory. Six men in an open boat risked life and limb and, if not smashed, sunk or lost, returned to process their prize by hand. There were no guarantees, and even the best technology made success a less than fifty-fifty proposition. Traditional whaling suggested a preindustrial charm and implied the whales had a chance.

Mechanised whaling, with spotter planes, catcher boats and factory ships, was a far cry from Ashley's and Melville's whaling. To the public, it was not a fair fight. As catches swelled in the 1950s and 1960s, populations crashed and the products returned, while valuable, were ultimately replaceable. The public's idea of whaling – an elemental challenge of man against beast where beast could prevail – failed to line up with Greenpeace's photographs. The Maritime Revival image of whaling had a far different effect on the public and policymakers from that of commercial fishing and resulted in the Marine Mammals Protection Act of 1972.

DISCOVERING THE SHORE

Romanticising preindustrial work in the machine age was only one way Americans redefined their interactions with marine environments. Seaside vacations grew in popularity with each passing decade after 1850. Options ranged from grand seaside hotels on New England's coast to more pedestrian boardwalks and Coney Island-style attractions. But before being lured to the shore, public perception of its desirability needed renovating. In antiquity, the sea was either a vast, enigmatic unknown or a threat causing harm via storms or an angry sea-god's wrath. Between 1750 and 1840, French historian Alain Corbin argues, artistic and literary Romanticism shifted the cultural meaning of beaches, harbours and seascapes. A growing sense of sublimity and sensuality, and new ways of seeing working ports and marine disasters, helped redefine westerners' relations with the shoreline (Corbin 1994, pp. 10–12, 73–97, 121–148, 187–198, 242–250). Corbin's nuanced argument holds up on this side of 'the Pond', even though redefining the shore took a few extra decades – perhaps resulting from America's comparative youth. By 1851, Herman Melville observed 'crowds of water-gazers ... fixed in ocean reveries' daily ringing his native Manhattan and his maxim 'as everyone knows, meditation and water are wedded for ever', drives home one of Corbin's most salient points – sublimity (Melville [1851] 2002, p. 19).

It was a fairly recent phenomenon, though – more of a shotgun wedding. Meditation and water had not been 'wedded for ever', in either Europe or America. The shore was a place of work, not play; or it was dangerous, with storms, or pimps, crimps (unscrupulous shipping agents and labour brokers) and prostitutes. However, once it was redefined as desirable and charming, human alterations of the coastal zone entered a new phase. Humans had adapted coasts as workplaces; now they would open them as venues of play, bringing vast new numbers to consume new pleasures.

Beachgoers wanted luxury, comfort and safety – all of which modified the physical environment. For five decades, from 1860 to 1911, W.H. Taylor operated a bath house and observed changes to life on the beach and boardwalk at Narragansett Pier, Rhode Island. Taylor recognised his beach's financial possibilities and offered patrons private dressing rooms, rented Victorian bathing suits and built a boardwalk over the sand.[10] 'There was no summer resorts at that time in the country', Taylor believed and soon competition from the Narragansett House attracted 'the very best families in the country', By 1870, Narragansett hoteliers had built over a dozen hotels ranging from fifty to over 200 rooms (Taylor 1927, p. 12) A McKim, Mead and White-designed pier and casino (1883–1886) also attracted visitors, making the resort a social centre.[11] As intended, 'it brought a great many yachts and this made a great many dinner parties and brought the fashionable class together to show their dresses and to be seen' (Taylor 1927, p. 12; Brown 1995, p. 203). Beyond social ebb and flow, Taylor was quick to note development's negative environmental effects, complaining the new jetty scoured sand from his beach (Taylor 1927, pp.13–15).

Resorts modified coastal zone geography, whether through docks to accommodate steamships at the massive Larkin House in Watch Hill, Rhode Island (1869, dock by 1889); boardwalks (Ocean City, MD, 1902; Atlantic City, NJ 1870, 1884); or even the hotel itself, as in San Francisco's third Cliff House (1896) perched upon the very edge of the living rock.[12] Moreover, transportation needs strained coastal resources, whether via trolley, railroad, auto road or steamer.[13]

SEAFOOD'S STATUS OR SEAFOOD AS STATUS?

Eating seafood unites fishing, maritime imagery in art and literature, public acceptance of the shore and seaside resort vacations. Consuming seafood was nothing new – Euro-Americans had relied on the sea's bounty since first contact – and Basque cod-fishing, Caribbean salt fish markets and feeding lobster to convicts and servants are old news. What the Maritime Revival changed, however, was the cultural meaning assigned to seafood, the types of food consumed and the venues where it was eaten. By the mid-1840s, salted fish lost its desirability and middle-class status through association

10 Taylor 1927, pp. 25–7. Taylor's spelling has been standardised. His grammar has not.

11 Narragansett, Rhode Island, Chamber of Commerce website.

12 Barnes 2005, pp. 47–8; Ocean City Boardwalk website; Casino Reinvestment Development Authority website; Blaisdell 2005.

13 Coney Island Timeline website. Coney Island's railroad opened in 1877 and the New Iron Pier opened in 1880 to accommodate steamships.

with urban poor, immigrants and southern blacks. Thanks to transportation infrastructure and refrigeration, fresh seafood gained popularity and status.[14]

Originally consumed primarily by those who caught it or lived near water, fish presents a conundrum, growing in status while at the same time being consumed by immigrants and urbanites. Tourism and a growing vacation industry help explain the shift. As city dwellers headed off to Vermont farms, Cape Cod and the islands or the White Mountains, they expected 'plain country fare'. Of course, this was not what country folk actually ate; instead, it was what guests imagined as food appropriate to the setting (Brown 1995, pp. 157–9). On farms, visitors wanted fresh butter, cream and eggs. Likewise, shore visitors expected seafood, best exemplified by lobster's move from being essentially undesirable in Maine to its current position as representative of the state itself.[15]

One event brought seafood to the great centre – gastronomically and geographically – of the United States. More than 27 million people gained admission to the 1893 World's Columbian Exposition at Chicago, a signal event attended by Americans across economic, racial and class lines (US Bureau of the Census 1991; Badger 1979, p. 109). Educating visitors was as important as entertainment, and the Fisheries Building held a prominent position at the fair (Rydell 1984, pp. 7, 241; Craig 2004, pp. 28, 30). Gloucester, Massachusetts held a similarly-prominent place inside the building, championing their industry and fishermen 'who spend their lives in the Banks of Newfoundland in hazardous and weary toil' (Massachusetts Board of World's Fair Managers 1894, p. 178). Between the building's exterior sculptures, enormous aquaria, theatrical lighting and the classification of some species as 'plebeian' and others 'terrors of the deep', one might surmise oceans were being mystified. Despite dramatic flourishes, the Fisheries Building actually demonstrated human mastery over nature by using heroic imagery alongside natural history.

Dining venues were nearby. The Café de la Marine offered patrons seafood, still uncommon for Midwesterners (Appelbaum 1980, p. 72). An eating saloon graced the Fisheries Building's upper floors, providing 'those who patronize fish dinners at the Exposition' examples of 'the value of fish as food'. It also reinforced nature as a commodified resource (Flinn 1893, pp. 59, 61). Finally, 'epicurean visitors' could partake of a 'continuous

14 O'Leary 1996, pp. 265–6, 270. See Hooker 1981, Oliver 1995 and Untermeyer 1932.

15 Conforti 2001, pp. 239, 353. See Lewis 1989.

clam-bake' in the nearby New England Clam-Bake Building (Bureau of Charities and Correction 1893, p. 15). The Exposition helped bring exotic ocean products to unfamiliar inland audiences, publicised products and industries to new consumers and set new standards of taste. The Fisheries Building and restaurants depicted nature as a force humans could conquer. With symbolism reminiscent of the Greco-Roman gods whose statues presided over the Exposition, there was no more powerful act than eating something to attain dominion over it.

CONCLUSIONS

As oceans were demystified and seafarers heroicised in popular culture, human mastery over nature appeared complete. Older working-class experiences were recast as authentic in a mechanised era, coastal real estate shifted from work zone to playground and the products of oceanic extraction elevated as delicacies for the wealthy – or those striving to be wealthy. Art, literature, advertising, public memorials and museums reinforced these new cultural meanings of the sea.

Seafaring's redefinition had direct consequences for policies and attitudes toward coastal zone usage. No longer devoid of substantial construction, vacation communities grew in places previously shunned as dangerous, cold or flood-prone, and development altered coastal geographies in ways ranging from wastewater treatment to erosion. As more people took to the shore for the day, week or season, their new tastes dictated everything about shore vacations from cottage decoration to recreation. Changing foodways put added pressure on ocean resources, resulting in increased fishing efforts. Coming full circle, fishing communities responded to changes brought by the Maritime Revival by claiming a birthright manufactured by the Revival itself.

The idea that seafaring is different from terrestrial work – more heroic, pure and preindustrial – offered upper-class vacationers and year-round coastal dwellers alike the opportunity to laud an imagined maritime past. For the former, it defined tastes and values and helped commodify material and intellectual culture. For the latter, it offered the cultural tools to sustain a lifestyle and its consequences by identifying with a preindustrial past, despite working in a modern age.

Escaping the Maritime Revival viewpoint is no easy task. Everyone – from policymakers, to participants, to the public – exists within its century-plus milieu. Moreover, enjoying its fruits, from ship models to shore dinners, increases chances of embracing a revivalist viewpoint uncritically. Still, some strategies allow enjoyment of beaches, paintings and seafood without losing one's critical lens. Recognise the past, especially the experience-as-lived, as different from a romantic perspective. Recover the rough edges excised while constructing the image. Consider the individuals or groups using sanitised, romantic or heroic imagery. Who has a stake? Hoteliers? Restaurateurs? Fishermen? Politicians? Realise some motivations are more sinister than others: a marine artist likely does less harm than a developer filling wetlands. Most importantly, consider oceans an extension of the terrestrial world. Though the sea is an alien environment seemingly invisible below the surface, humans should not need SCUBA gear to realise that what happens on land has effects everywhere; and no amount of heroic art or literature will eliminate the consequences of human activities on the marine environment.

BIBLIOGRAPHY

Appelbaum, S. 1980. *The Chicago World's Fair of 1893: A Photographic Record.* New York: Dover.

Ashley, C. 1926. *The Yankee Whaler*. Boston: Houghton Mifflin.

Axelrod, A. (ed.) 1985. *The Colonial Revival in America.* New York: Norton/Winterthur Museum.

Babson, J.J. 1972 [1860]. *History of Gloucester, Cape Ann, Including the Town of Rockport.* Gloucester, MA: Peter Smith.

Badger, R. 1979. *The Great American Fair: The World's Columbian Exposition and American Culture.* Chicago: N. Hall.

Barnes, C.B. 2005. *Watch Hill Through Time: The Evolution of a New England Shore Community.* Watch Hill, RI: Watch Hill Conservancy.

Beaulieu, F. 22 July 2010. 'Destructive NOAA, NMFS, Hide the Ocean Reality'. *Gloucester* (MA) *Times.* http://www.gloucestertimes.com/opinion/x1527083941/Destructive-NOAA-NMFS-hide-the-ocean-reality [Accessed 15 August 2011].

Berrill, M. 1997. *The Plundered Seas: Can the World's Fish Be Saved?* Vancouver, BC: Greystone.

Blaisdell, M. 2005. *San Francisciana: Photographs of the Cliff House.* San Francisco: Marilyn Blaisdell.

Brown, D. 1995. *Inventing New England: Regional Tourism in the Nineteenth Century.* Washington: Smithsonian Institution.

Casino Reinvestment Development Authority. 'Atlantic City Boardwalk: a Stroll on the Wooden Way is Steeped in History'. http://www.atlanticcitynj.com/mediaonline/story_ideas_detail.aspx?NewsID=80 [Accessed 9 September 2011].

Conathan, M. 2001. 'Fish on Fridays: What I Did on My Summer Vacation'. *Center for American Progress*. http://www.americanprogress.org/issues/2011/08/fof_081211.html [Accessed 15 August 2011].

'Coney Island Timeline'. http://www.westland.net/coneyisland/time.html [Accessed 5 September 2011].

Conforti, J.A. 2001. *Imagining New England: Explorations of Regional Identity From the Pilgrims to the Mid-Twentieth Century*. Chapel Hill: University of North Carolina Press.

Cooney, R.B. 1950. *Westerly's Oldest Witness: How Westerly and the Washington Trust Company Have Progressed Together for 150 Years*. Westerly, RI: Washington Trust Company.

Corbin, A. 1994. *The Lure of the Sea: The Discovery of the Seaside in the Western World, 1750–1840*. (Trans. Jocelyn Phelps) New York: Penguin.

Craig, W.O. 2004. *Around the World with the Smithsonian*. Washington, D.C.: Smithsonian Institution.

Denenberg, T.A. 2003. *Wallace Nutting and the Invention of Old America*. New Haven: Yale.

Ellis, R. 1991. *Whales and Men*. New York: Knopf.

Flinn, J.J. (compiler). 1893. *Official Guide to the World's Columbian Exposition*. Chicago: Columbian Guide Company.

Gabbatt, A. 12 December 2010. 'Pink Floyd Guitarist Son Charlie Gilmour Held Over Student Protests'. *The Guardian*. http://www.guardian.co.uk/education/2010/dec/12/charlie-gilmour-arrest-student-protests [Accessed 24 August 2011].

Gigante, D. 2005. *Taste: A Literary History*. New Haven: Yale University Press.

Gombrich, E.H. 2002. *The Preference for the Primitive: Episodes in the History of Western Taste and Art*. London: Phaidon.

Grasso, G.M. 2008. 'What Appeared Limitless Plenty: The Rise and Fall of the Nineteenth-Century Atlantic Halibut Fishery'. *Environmental History* **13**: 66–91.

— 2009. *The Maritime Revival: Antimodernity, Class, and Culture, 1870–1940*. Ph.D. diss. Durham, NH: University of New Hampshire.

Jackson, J.B.C. 2001. 'What Was Natural in the Coastal Oceans?' *Proceedings of the National Academy of Sciences* **98**: 5411–18.

Jackson, J.B.C. et al. 2001. 'Historical Overfishing and the Recent Collapse of Coastal Ecosystems'. *Science* **293**: 629–37.

Hooker, R.J. 1981. *Food and Drink in America: A History*. New York: Bobbs-Merrill Company.

Kerry, J. 8 March 2011. 'At Hearing, Kerry Drills Feds on Massachusetts Fishing Industry'. http://kerry.senate.gov/press/release/?id=cace359d-ca49-42c1-8033-ec69bb50946d [Accessed 10 September 2011].

Lears, T.J.J. 1981. *No Place of Grace, Antimodernism and the Transformation of American Culture, 1880–1920*. New York: Pantheon.

Leavitt, J.F. 1973. *The Charles W. Morgan*. Mystic, CT: Marine Historical Association.

Lewis, G.H. 1989. 'The Maine Lobster as Regional Icon: Competing Images Over Time and Social Class'. *Food and Foodways* **3**: 303–16.

Lindgren, J. 1999. '"Let Us Idealize Old Types of Manhood": The New Bedford Whaling Museum, 1903–1941'. *New England Quarterly* 72: 163–206.

Lynes, Russell. 1983 [1954]. *The Tastemakers*. Westport, CT: Greenwood.

Massachusetts Board of World's Fair Managers. 1894. *Report of the Massachusetts Board of World's Fair Managers*. Boston: Wright and Potter Printing.

McClenachan, L. 2009. 'Documenting Loss of Large Trophy Fish from Florida Keys with Historical Photographs'. *Conservation Biology* **23**: 636–43.

McEvoy, A.F. 1986. *The Fisherman's Problem: Ecology and the Law in the California Fisheries, 1850–1980*. Cambridge: Cambridge University Press.

Melville, H. 2002 [1851]. *Moby-Dick*. New York: Norton.

Morison, S.E. 1921. *The Maritime History of Massachusetts, 1783–1860*. Boston: Houghton Mifflin.

Narragansett, Rhode Island Chamber of Commerce. 'The Historic Narragansett Towers'. http://www.narragansettri.com/thetowers/index.htm [Accessed 3 September 2011].

North Shore Community College. 'The Poetry of Places in Essex County: The Fishermen's Memorial and The Fishermen's Wives Memorial (Gloucester Harbor) Location, History, and legends'. http://myweb.northshore.edu/users/ccarlsen/poetry/gloucester/memorialshistory.htm [Accessed 25 August 2011].

'Ocean City Boardwalk'. http://www.ocboards.com [Accessed 9 September 2011].

O'Leary, W.M. 1996. *Maine Sea Fisheries: The Rise and Fall of a Native Industry, 1830–1890*. Boston: Northeastern University Press.

Oliver, S. 1995. *Saltwater Foodways: New Englanders and Their Food, at Sea and Ashore in the Nineteenth Century*. Mystic, CT.: Mystic Seaport Museum.

Pauly, D. 1995. 'Anecdotes and the Shifting Baseline Syndrome of Fisheries'. *Trends in Ecology and Evolution* **10**: 430.

Procter, G.H. 1873. *The Fisherman's Memorial and Record Book*. Gloucester, MA: Procter Brothers.

Procter Brothers. 1882. *The Fishermen's Own Book*. Gloucester, MA: Procter Brothers.

Rhoads, W.B. 1977. *The Colonial Revival*. New York: Garland.

Roberts, C. 2007. *The Unnatural History of the Sea*. Washington, D.C.: Island Press.

Rocky Neck Historic Art Trail, 'A.W. Buhler Studio'. http://www.rockyneckartcolony.org/art_trail/Trail%20Site%205_Buhler.pdf [Accessed 25 August 2011].

Rosenberg, A. et al. 2005. 'The History of Ocean Resources: Modeling Cod Biomass Using Historical Records'. *Frontiers in Ecology and the Environment* **3**: 84–90.

Rydell, R.W. 1984. *All the World's a Fair: Visions of Empire at American International Expositions, 1876–1916*. Chicago: University of Chicago Press.

Santos, M.W. 2002. *Caught in Irons: North Atlantic Fishermen in the Last Days of Sail*. Snelinsgrove, PA: Susquehanna University Press.

Taylor, J.E., III. 1999. 'Burning the Candle at Both Ends: Historicizing Overfishing in Oregon's Nineteenth-Century Salmon Fisheries'. *Environmental History* 4: 54–79.

Urbon, S. 13 January 2011. 'Barney Frank Barks at Obama Administration Over Fishing'. *Boston Herald.* http://news.bostonherald.com/news/politics/view/20110113barney_frank_barks_at_obama_administration_over_fishing/ [Accessed 10 September 2011].

Untermeyer, L. 1932. *Food and Drink.* New York: Harcourt, Brace, and Company.

US Bureau of the Census. 1991. 'Statistical Abstract of the United States'. http://www2.census.gov/prod2/statcomp/documents/1991-02.pdf [Accessed 4 October 2008].

Vickers, D. 1994. *Farmers and Fishermen, Two Centuries of Work in Essex County, Massachusetts, 1630–1830.* Chapel Hill: University of North Carolina Press.

Vickers, D., with V. Walsh. 2005. *Young Men and the Sea: Yankee Seafarers in the Age of Sail.* New Haven: Yale, 2005.

Weaver, R. 1921. *Herman Melville: Mariner and Mystic.* New York: George H. Doran.

Manuscript Materials

Bureau of Charities and Correction, Circular No. 6. 'World's Fair Notes', bound into Davis and Skiff. 1893. *Classification of Department of Transportation Exhibits: Railways, Vessels, Vehicles, with other information for intending exhibitors.* Chicago: Donahue and Henneberry. Boston Athenaeum Collection.

Typescript, W.H. Taylor. Circa 1927. 'Memories of Narragansett Pier, Part II'. David Patten Papers, MSS 605. Rhode Island Historical Society Collection.

~4~

Beneath the Surface of the Pacific Waters: John Steinbeck's (Deep) Ecological Explorations

Petr Kopecký

INTRODUCTION

While, in the minds of most readers, John Steinbeck still figures as a socially-engaged writer, the environmentalist dimension of his work has remained rather obscure. This is all the more surprising given the fact that 'the one book he wished he had written' was *Silent Spring* (1962) by Rachel Carson (Shillinglaw 2011), as Steinbeck admitted toward the close of his life. At approximately the same time he confessed that, out of his own writing, the book he was most proud of was the narrative portion of *Sea of Cortez* (1941), which was published in 1951 as *The Log from the Sea of Cortez* (abbreviated here as *LSC*).[1] With hindsight, this particular narrative arguably had potential to raise ecological awareness like Carson's books.[2] However, its critical reception was muted by the events in the Central North Pacific which had captured the attention of the nation.

This essay explores the multi-faceted nature of *LSC*. The book is a crucial source of ideas that shaped Steinbeck's worldview and eventually found their way into his best-known works. However, at the time of the book's publication, few people realised its pioneering quality. Among these few was Joseph Campbell, who later became a renowned mythologist.[3] In

1 *LSC* is the narrative portion of *Sea of Cortez*. The original 1941 *Sea of Cortez* also includes a detailed catalogue of species. Although Steinbeck is listed as the sole author of *LSC*, largely owing to the marketing interests of the publisher, we can now safely say that Ed Ricketts substantially contributed to the narrative.

2 Besides *Sea of Cortez*, late fall 1941 also saw the publication of Carson's *Under the Sea-Wind* which also crosses and/or dissolves the boundaries between biological classes and groups.

3 Campbell befriended Steinbeck and Ricketts during his prolonged stay in Monterey in 1932. Many of their conversations centred around the manuscript of what was to become *To a God Unknown*.

his letter to Ricketts, Campbell noted that 'it is a great book, dream-like and with no end of implications, sound implications!' (quoted in Larsen and Larsen 1991, p. 203). Yet those sound implications were not detected by the majority of literary critics for a long time.

The critical reception of the narrative at the time of its publication was somewhat bewildered. At that time – and not only when assessing Steinbeck's work – critics were neither ready nor qualified to appreciate the environmentalist (as opposed to humanist) agenda that permeates the narrative. A few reviewers, most importantly Joel W. Hedgpeth (1953), could at least recognise the ecological nature of the author's observations of the physical environment of the littoral. However, they failed to appreciate the significance of the metaphysical extensions of these observations. A serious exploration of Steinbeck's metaphysical contemplations came with the emergence of ecocriticism, which concentrates on the 'green' aspects of literary works. This essay thus examines the interface between the science of ecology and its metaphysical extension, known as deep ecology, in Steinbeck's writing, especially in *LSC*. It should be noted that the ecological dimension of Steinbeck's writing has received considerable attention, whereas its deep ecological elements have occasionally been identified by scholars, but never adequately probed. Although the boundary between ecology and deep ecology is permeable, not only in Steinbeck's work, I deal with them separately.

ECOLOGICAL FOUNDATIONS

Two acclaimed literary scholars, Jackson J. Benson (1984, p. 244) and Robert DeMott (1979: pp. 86–9), have persuasively contended that the impact of science on Steinbeck's writing was unparalleled among the major US fiction writers of his time. As has been suggested above, this ecological dimension did not win acclaim for Steinbeck in critical circles. On the contrary, as Benson argues, the ecological view of the world was a novelty even to biological scientists when the narrative was first published in 1941, 'and as far as the literary world was concerned, it could have been a Martian religion. He didn't realise – nor would he have cared, if he had – that he was on the road to offending nearly everyone.' (Benson 1984, p. 431) It is important to note that the US involvement in WWII accelerated progress in industry, which in turn required even greater specialisation in scientific research. This development inevitably led to increasing reductionism on the scientific front.

Steinbeck's understanding of ecology was bolstered by his closest friend, marine biologist Ed Ricketts. Nevertheless, the author's interest in ecological sciences predates his meeting with Ricketts. During his somewhat leisurely studies at Stanford University, Steinbeck took a course in zoology with a focus on marine biology. The course in question was taught at Hopkins Marine Station in Monterey in 1923. It introduced Steinbeck to William Emerson Ritter's organismal theory, which considerably influenced the writer's view of nature. It was again in Monterey, in the 1930s, that the seeds sown while at university germinated and ripened in Steinbeck's informal but all the more intense discussions with Ricketts, who managed the Pacific Biological Laboratory on what is today known as Cannery Row. To Steinbeck, Ricketts was a great source of inspiration as well as scientific information. He was also the model for some of the most memorable characters in Steinbeck's fiction, including Doc Burton in *In Dubious Battle* (1936), Jim Casy in *The Grapes of Wrath* (1939) and Doc in *Cannery Row* (1945).

In his lifetime, Steinbeck was often read and interpreted through the lens of political and social theories. While it is true that he was informed by the theories of Herbert Spencer and William McDougall, who conceived of society as an organism,[4] Steinbeck's conception of man and/in society owes more to pioneering ecologists, especially the aforementioned William Emerson Ritter and Warder Clyde Allee. Regarding Ritter's organismal theory, Steinbeck was particularly interested in the concept of superorganism. The whole (the superorganism) comprises parts: 'wholes are so related to their parts that not only does the existence of the whole depend on the orderly cooperation and interdependence of its parts, but the whole exercises a measure of determinative control over its parts' (quoted in Hedgpeth 1971, p. 95). Ritter's theory complemented the ideas of Allee, who taught Ed Ricketts at the University of Chicago in the 1920s.[5] Ricketts, and eventually also Steinbeck, drew on Allee's analysis of animal aggregations and the related theories of mutual interdependence and automatic cooperation. Ed Ricketts's laboratory witnessed countless debates between Steinbeck and Ricketts on the principles postulated by Ritter and Alee and their implications.

4 For more on Herbert and McDougall, see Fontenrose 1964.

5 The complementary nature of the ideas of Ritter and Allee is discussed by Richard Astro (1995, pp. vii–xxiii).

These progressive ideas found their way into the works published by both men in April 1939, namely Steinbeck's *The Grapes of Wrath* and Ricketts's *Between Pacific Tides* (co-authored by Jack Calvin). While Steinbeck's masterpiece immediately captured the attention of Americans, Ricketts's and Calvin's book was not fully appreciated until much later. Part of the reason is that Ricketts was not well respected in the community of scholars because he had no academic degree. In addition, his ecological approach to the study of the littoral was ahead of time. In the words of John Pearse, Professor Emeritus of ecology and evolutionary biology at UC Santa Cruz, 'it wasn't until the 1980s that scientists started to look at the intertidal the way Ricketts did' (quoted in Tamm 2005, p. E 2.) According to Eric Enno Tamm (2004, p. 33), *Between Pacific Tides* 'is one of the best-selling books ever published by Stanford University Press, selling some 100,000 copies'.

The ecological pattern is perhaps most conspicuous in another novel centred around the Rickettsian character Doc, namely *Cannery Row* (1945). In this book, Steinbeck weaves an ecological web out of the diverse characters and stresses the commensal spirit of the neighbourhood. The life of the community on Cannery Row is strikingly similar to that in a tide pool. The tide pool is also deployed as a central metaphor in *LSC*. It lucidly manifests the cyclical nature of life. Thus, it can be poetically rendered as a tomb and womb at the same time.

Katharine A. Rodger regards *LSC* as a 'seminal work on ecological holism' (Rodger 2002, p. xii). Its originality lies in the leisurely nature of the narrative (the subtitle of *Sea of Cortez* is *A Leisurely Journal of Travel and Research*), which allowed the author to do something that was then hardly acceptable in the academic community, namely to apply a cross-disciplinary approach. In *LSC*, Steinbeck smoothly crosses what he sees as porous boundaries between different areas of study. In doing so, he contemplates the role of science. He repeatedly invokes Charles Darwin and his inductive reasoning, and laments that modern scientists' method all too often lies in 'quickly looking at the whole field and then diving down to a particular' (*LSC*, p. 70). Steinbeck also finds fault with the narrowing horizons of contemporary science, which was becoming too specialised and readily served the needs of the economy and industry. He expresses dissatisfaction with the kind of natural science that accumulates vast amounts of data but loses sight of the fundamental unity of life: 'The process of gathering of knowledge does not lead to knowing' (*LSC*, pp. 196–7). His overall concern is the growing atomisation of science, and his remedy to this undesired development is ecological holism. To make his point in *LSC*, Steinbeck uses the tide pool metaphor to demonstrate the need for a broadening of man's vision: 'If he has strength and energy of mind the tide pool stretches both ways, digs back to electrons and leaps space into the universe and fights out the moment into non-conceptual time. Then ecology has a synonym which is ALL.' (*LSC*, p. 99) As in many other passages throughout the book, he casts light on the interconnectedness between the invisible life units in the tide pool and the incomprehensible forces.

Steinbeck also points out the importance of the superorganism – which bears striking resemblance to the ecosystem, a term that was coined in 1935 but came into general use much later. The research conducted by Steinbeck and Ricketts on the littoral, as well as the way it is conveyed in the narrative, deviated from the standard taxonomic approach of the time which was preoccupied with individual species rather than the interaction of organisms with their environment:

> [N]ot only the meaning but the feeling about species grows misty. One merges into another, groups melt into ecological groups until the time when what we know as life meets and enters what we think of as non-life: barnacle and rock, rock and earth, earth and tree, tree and rain and air. And the units nestle into the whole and are inseparable from it. (*LSC*, p. 257)

By the standards of the time, this perspective was certainly innovative.

While it is evident that *LSC* is deficient in specialised terminology and could also be challenged on the grounds of scientific accuracy, its strength lies in its holistic approach and the author's determination to uncover the common roots of the rapidly growing separate branches of natural sciences. Thus his endeavour belongs to the tradition of Darwinian naturalism rather than modern science. Steinbeck also aimed to communicate his findings in a manner accessible to lay readers. In so doing, he sometimes resorted to untenable analogies and simplifications. Nevertheless, the holistic outlook allowed him to predict what others failed to see, especially the drastic reduction in the sardine population as a result of overfishing (also discussed in *Cannery Row* and *Sweet Thursday*). That Steinbeck and Ricketts made a significant contribution to science is also attested by the species found by and/or named after them (e.g. *Thalassema steinbecki*, *Phialoba steinbecki*, *Siphonides rickettsi*, *Isometridium rickettsi* and many others).[6]

DEEP ECOLOGICAL MEDITATIONS

While the scientific insights in *LSC* are generally prescient, often stemming from the latest developments in the field of ecology, the real cutting-edge quality of the work can be found in the passages where Steinbeck breaks away from the strictures of science-neuter. In a 1941 letter to his editor, Steinbeck writes about the careful plan for the book, which consists of 'four levels of statement', adding that 'few will follow it to the fourth. I even think that is a new kind of writing. I told you once that I found great poetry in scientific writing' (Steinbeck and Wallsten 1976, p. 232). The deeper levels lie in the parts devoted to scientific pursuits. These parts of the book harbour ecophilosophical ideas which were later to become the cornerstones of one of the most radical streams of modern environmentalism, namely deep ecology. The term 'deep ecology' was coined by the pre-eminent Norwegian philosopher Arne Naess in 1972. It soon found fertile ground in the United States, especially in California, where two scholars, Bill Devall and George Sessions, published a seminal text of deep ecology titled *Deep Ecology: Living as if Nature Mattered* (1984). The authors acknowledge the importance of literature as a vehicle that can effectively convey deep ecological concepts. They include and comment on the works of several American authors (John Muir, Mary Austin, Robinson Jeffers, Gary Snyder, etc.). Steinbeck was beyond their radar. Nevertheless, in private correspondence, Bill Devall admitted in 2004 that *LSC* was relevant to deep ecology and 'not including Steinbeck [in *Deep Ecology*] was a sin of omission' (Devall 2004).

The pivotal idea of deep ecology is non-anthropocentrism. The essential tenet of the deep ecology platform concerns the intrinsic value of nonhuman life forms. In *LSC* Steinbeck breaks down the foundations of the hierarchical conception of life, the ancient chain of being, and replaces it with an egalitarian one. In his view, grounded in scientific holism, humans stand neither above nor next to nature: they are an integral part of it. Steinbeck persistently challenges the general belief that '[M]an grows toward perfection; animals grow toward man' (*LSC*, p. 100). Instead, he

6 For the complete list of species named after Ricketts and Steinbeck, see Tamm 2004, p. 243.

emphasises the animal-like nature of people, to whom he often refers as a species, and asks provocatively: 'Why do we so dread to think of our species as a species? Can it be that we are afraid of what we may find?' (*LSC*, p. 314) As in his fiction, in *LSC* Steinbeck sometimes anthropomorphises animals and animalises people. While this literary device is often classified as pathetic fallacy, it can also be interpreted as a means of acknowledging the equal status of people and animals.[7]

The most powerful passages exemplifying what ecophilosophers would later call biological egalitarianism are also built around the littoral. Indeed, Steinbeck realises that the tide pool is a potent metaphor of the permeable borderline between land and sea. He aptly applies the notion of non-divisibility and elevates it to an ontological level:

> [M]an is related to the whole thing, related inextricably to the whole reality known as unknowable. This is a simple thing to say, but the profound feeling of it made a Jesus, a St. Augustine, a St. Francis, a Roger Bacon, a Charles Darwin, and an Einstein. Each of them in his own tempo and with his own voice discovered and reaffirmed with astonishment the knowledge that all things are one thing and that one thing is all things – plankton, a shimmering phosphorescence on the sea and the spinning planets and an expanding universe, all bound together by the elastic string of time. It is advisable to look from the tide pool to the stars and then back to the tide pool again. (*LSC*, p. 257)

Steinbeck also harbingered what is known today as sustainability, a current buzzword among modern environmentalists as well as business people. He urgently felt the necessity of 'a continuing economy'. This appeal stems from the author's awareness of the fact that the planet's resources are finite, which he voiced most clearly in *Sweet Thursday* and *America and Americans*. These ideas were to become the basic assumptions of sustainability.

In *LSC*, Steinbeck applies this model to fishing, calling for much stricter catch limits. Toward the end of his life, he shifted closer toward the utilitarian rhetoric of Wise Use. In a remarkable article titled 'Let's Go After the Neglected Treasures Beneath the Seas', which was printed in *Popular Science* in 1966, he writes that '[w]e must study, control, herd, and improve the breeds of animals, because we are shortly going to need them' (Steinbeck 1966a, pp. 84–7). Within this single article, Steinbeck demonstrates his inconsistency, on the one hand stressing the dwindling resources, while on

7 See, for instance, Lakoff and Turner 1989, p. 72.

the other hand identifying plankton as a 'boundless bank of protein food' (p. 86). Nevertheless, in the same article he formulated several propositions that presupposed a radical reconsideration of the predominant policies of the day.

FINAL FRONTIER AND NEW HORIZONS

The theme of the final frontier plays an important role in Steinbeck's work. One does not have to artificially impose the theme on the work; it pervades many of his novels, such as *To a God Unknown*, *The Grapes of Wrath* and *East of Eden*. In more or less subtle ways, Steinbeck conceives of the California coastline as the final frontier, both geographical and psychological. It is the terminal station of the westward movement through which America and Americans were dynamically evolving when the dream of Eden still remained alive. The closing of the American frontier in the 1890s resulted in the incapacitation and restlessness of people who had been constantly in flux. The most eloquent manifestation of this idea comes in *The Red Pony* (1933): 'The westering was as big as God, and the slow steps that made the movement piled up and piled up until the continent was crossed. "Then we came down to the sea and it was done … There's no place to go. There's the ocean to stop you. There's a line of old men along the shore hating the ocean because it stopped them".' (Steinbeck 2000 [1933], pp. 95–6)

The ocean is not hated just because it is a territory that can hardly be tamed and domesticated. It represents an insurmountable geographical barrier and, even more importantly, it is a psychological barrier. It is not a coincidence that Steinbeck looked at the ocean through the lens of depth psychology.[8] He was particularly influenced by the Jungian hypothesis of the collective unconscious. In Steinbeck's books, the ocean is associated with what is yet beyond human control and understanding, the dark sphere of the unknown. Simply stated, the ocean acquires an archetypal quality in Steinbeck's work. Once again, this trait is best defined in *LSC*:

> For the ocean, deep and black in the depths, is like the low dark levels of our minds in which the dream symbols incubate … We have thought often of sea-memory, or sea-thought, which lives deep in the mind. If one ask for the description of the unconscious, even the answer-symbol will usu-

8 In this respect, Steinbeck's work was informed by Carl Gustav Jung and, above all, his acquaintance from Monterey Joseph Campbell.

> ally be in terms of a dark water into which the light descends only a short distance. (*LSC*, pp. 36–7)

Nevertheless, Steinbeck offers a much wider spectrum of responses to the final frontier in his work. While the encounter with the ocean represented closure for the 'line of old men' in *The Red Pony*, to others it was an opening of new horizons, a new challenge. Steinbeck vaguely formulated this notion as early as 1929, noting that '[T]he new eye is being opened here in the west – the new seeing' (quoted in Valjean 1975, p. 123).

Arguably, this new seeing stood for Steinbeck's emerging ecological vision, which was then being postulated through *To a God Unknown*. It stands in opposition to the frontier mentality of the people for whom the bountiful natural resources of America appeared inexhaustible. The new seeing, in the form of environmental consciousness, began to arise with the realisation of the finality of the seemingly endless continent, a realisation that occurred on the Pacific shores. The writer anticipated and perhaps even accelerated the coming of the environmental wave that swept across the country in the 1960s. He saw the astounding diversity and complexity of the ocean and its interdependence with terrestrial life. He saw that the ocean needed both exploration and preservation. But he also foresaw that the ocean might be just another frontier conquered by man.

CONCLUSION

The preceding pages have shown Steinbeck as a largely 'green' writer. The seventy plus years that have passed since the publication of *Sea of Cortez* have witnessed great advances in various scientific fields. It is especially through their cross-pollination that many of Steinbeck's ideas have re-entered the academic as well as the public arena. Nowadays, people have a much better understanding of Steinbeck's observations of the environment and his related philosophical insights. Steinbeck himself foretold the uneasy reception of the book, suggesting that 'gradually it will be discovered that it is a whole new approach to thinking and only very gradually will the philosophic basis emerge' (Steinbeck and Wallsten 1976, p. 230). This prediction has turned out to be very accurate. It took some time before literary critics were able to decode the holistic design of this narrative (and that of his other books), and it took a whole new current of literary studies, ecocriticism, to grasp the philosophical foundations upon which the narrative is built.

Ecocritical analysis of Steinbeck's life and work has uncovered an ecological design behind some of his best writing. It also shows the holistic mindset of the author, which may well be attributed to his Romantic leanings. Both consciously and subconsciously, Steinbeck responded to the ongoing fragmentation of the society by searching for connections and unity. No wonder he deployed generic techniques that were more closely linked with Romantic literature, such as fables, pastoral and allegory. These well served his ultimate concern, which was to present an organic and inclusive picture of the interactions between the human and nonhuman world.

ACKNOWLEDGEMENTS

The financial support of the Czech Science Foundation (GA ČR), Grant No. P406/10/P297 ('Jeffersova a(nebo) Steinbeckova země: literární bitva o Kalifornii') is gratefully acknowledged.

BIBLIOGRAPHY

Astro, R. 1995. Introduction to *The Log from the Sea of Cortez*. New York: Penguin Books, pp. vii–xxiii.

Benson, J.J. 1984. *The True Adventures of John Steinbeck, Writer*. New York: The Viking Press.

Carson, R. 2007. *Under the Sea-Wind*. New York: Penguin Books.

DeMott, R. 1979. 'The Interior Distances of John Steinbeck'. *Steinbeck Quarterly* 7 (3–4): 86–99.

Devall, B. 2004. 'Re: California landscapes and origins of deep ecology'. E-mail to Petr Kopecký. 1 April 2004.

Fontenrose, J. 1964. *John Steinbeck: An Introduction and Interpretation*. New York: Barnes and Noble.

Hedgpeth, J.W. 1971. 'Philosophy on Cannery Row', in R. Astro and T. Hayashi (eds) *Steinbeck: The Man and his Work* (Corvallis: Oregon State University Press) pp. 89–129.

— (1953) '*Sea of Cortez* Revisited, or *Cannery Row* Revised'. *Pacific Discovery* 6 January–February: 28–30.

Lakoff, G. and M. Turner. 1989. *More than Cool Reason: A Field Guide to Poetic Metaphor*. Chicago: University of Chicago Press.

Larsen, S. and R. Larsen. 1991. *A Fire in the Mind: The Life of Joseph Campbell*. New York: Doubleday.

Ricketts, E. and J. Calvin. 1939. *Between Pacific Tides*. Stanford: Stanford University Press.

Rodger, K.A. 2002. *Renaissance Man of Cannery Row: The Life and Letters of Edward F. Ricketts*. Tuscaloosa and London: The University of Alabama Press.

Shillinglaw, S. 'The Steinbeck Institute 2011 – Dear Colleague Letter'. 15 April 2011. John Steinbeck: Voice of a Region, Voice for America. http://www.steinbeckinstitute.org/letter.html

Steinbeck, J. 1966. *America and Americans*. New York: Viking Press.

— 1966a. 'Let's Go After the Neglected Treasures Beneath the Seas'. *Popular Science* **189**, September: 84–87.

— 2000 [1933]. *The Red Pony*. New York: Penguin Books.

— 2006 [1936] *In Dubious Battle*. New York: Penguin Classics.

— 1992 [1939]. *The Grapes of Wrath*. New York: Penguin Classics.

— 1986 [1941]. *The Log from the Sea of Cortez*. New York: Penguin Books.

— 1994 [1945]. *Cannery Row*. New York: Penguin Books.

— 1996 [1954]. *Sweet Thursday*. New York: Penguin Books.

Steinbeck, E. and R. Wallsten (eds). 1976. *Steinbeck: A Life in Letters*. New York: Penguin.

Tamm, E.E. 2004. *Beyond the Outer Shores: The Untold Odyssey of Ed Ricketts, the Pioneering Ecologist Who Inspired John Steinbeck and Joseph Campbell*. New York and London: Four Walls Eight Windows.

— 2005. 'Of myths and men in Monterey'. *San Francisco Chronicle* (16 October 2005): p. E2.

Valjean, N. 1975. *John Steinbeck the Errant Knight: An Intimate Biography of his California Years*. San Francisco: Chronicle Books.

II. HISTORICISING SEAS AND MARINE LIFE

~5~

To the Islands: Ecological Imperialism on the North-West Australian Coast

Joseph Christensen

Among the mysteries of Australia's colonial history is the fate of the Yinikutira, an Aboriginal society that inhabited the North West Cape region on the continent's North-west coast. There is only one written account of their daily life and culture, a particularly rich ethnographic record produced from the testimonies of two European castaways shipwrecked on the Cape in 1875 (see Skurla 2009). It captures the Yinikutira in the early stages of their first sustained contact with an outside civilisation, where a lifestyle based around fishing and the hunting of turtles and dugongs was exposing the tribe to the presence of a commercial pearling industry that was developing rapidly in the North-west. The eventual saviour of the castaways, the colonial mariner Charles Tuckey, was one of the prominent pearling masters during the emergence of this industry. What became of the Yinikutira after 1875 remains unknown. When European pastoralists moved into the Cape region in the late 1880s they found a land that had been abandoned by its traditional inhabitants (Dench 2009, pp. 139–50).

Two explanations can be advanced for the disappearance of the Yinikutira. In 1875 the North West Cape lay astride the main pearling grounds of Shark Bay, to the south, and the North-west pearling grounds that stretched away to the east, and like other Aboriginal tribes in the North-west their fate is almost certainly tied up with the devastation visited upon this coast by the rise of pearling: the Yinikutira may have been coerced into labour aboard the pearling fleets through the practise known as 'blackbirding', a form of mass kidnapping that was commonplace in the pearling industry in the remote and largely lawless North-west; or they may have been struck by one of the many epidemics of exotic disease that passed through the indigenous population of this region following the arrival of European settlers (Dench 2009, pp. 139–50). Survivors from either of these scenarios are likely to have merged into the Aboriginal communities that formed on the fringes of white settlements during the late nineteenth century, where their identity as a people would have faded over time.

In short, the Yinikutira appear to have been victims of the spread of European imperialism. Yet, whereas British colonisation in Australia heralded the widespread destruction of Aboriginal civilisation through the combined effects of land dispossession, violence and disease, the contact situation in the North-west has fundamental differences to that in other parts of the continent. Between 1863, when the first settler landed in the Pilbara district with six cattle and 540 sheep, and 1913, when 116,858 cattle and 1,528,831 sheep were spread across 47,974,665 acres of pastoral leases, this process was largely complete (Battye 1915, p. 52). The population of the 24 language groups that existed at the time of first contact had been greatly reduced (precise figures are unavailable) and the remaining population was confined largely to a pauperised existence on the margins of settler society (Hasluck 1942, pp. 18–41). What stands the North-west apart from other Australian settlement frontiers is the impact of maritime activity in this conquest of the land. Pearling, through its demands for labour and its role in spreading disease, had been the major driver of the dispossession and depopulation of Aboriginal society across this region.

The aim of this chapter is to investigate the ecology of imperialism in the North-west. Alfred Crosby's *Ecological Imperialism* (1986) attributes the success of colonisation in the Americas and Australasia to the biological advantages that European plants, animals and pathogens gave to the settlers of temperate regions in these lands. In Crosby's analysis the sea is the medium that facilitates the creation of 'neo-Europes' across the oceans; mastery over navigation on the high seas provides a pathway for the biological transfers that underpins European expansion in the colonial era (Crosby 1986, pp. 1–8). North-west Australia is different, not only because its environment of a semi-arid continental hinterland and a sub-tropical littoral interspersed with hundreds of small islands is so alien to the temperate regions where European colonists normally prospered, but also because this seaboard was more than simply a medium for the advance of imperialism. Instead, through the pearling industry that it supported, the sea was itself an agent of imperialism, conditioned principally by the industry's bathymetric limit, or the depth at which pearlshell could be obtained on the seafloor.

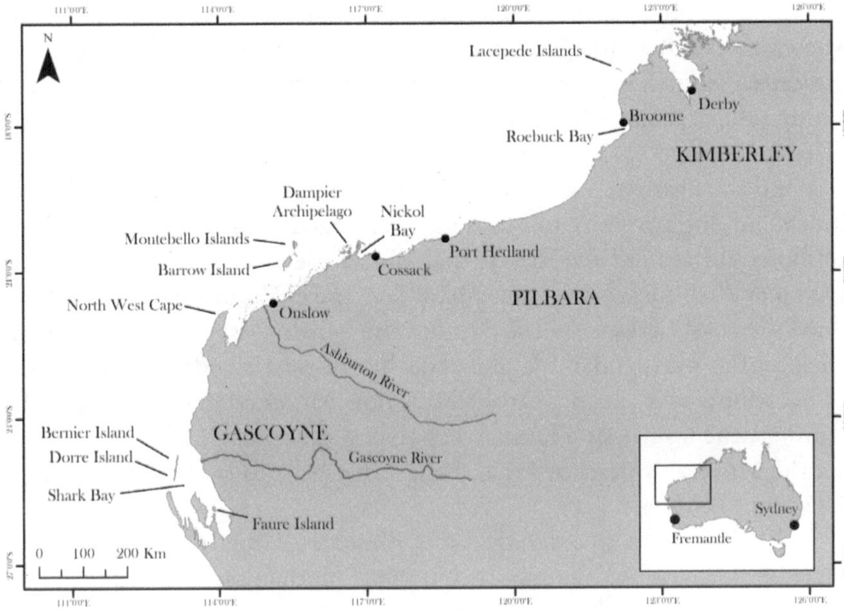

Map 1: The North-west Australian coast.

Pearling in Australia began at Shark Bay in the 1850s, based on the small oriental or Shark Bay pearl oyster *Pinctada albina*. The industry took off further along the North-west coast around 1866 when the first European settlers, newly arrived in the Pilbara, began to collect the large Mother-of-pearl shell *Pinctada maxima* as a sideline to pastoralism. It proved to be a profitable activity. Markets for Mother-of-pearl existed in Britain and Europe, where pearlshell was manufactured into buttons, inlays on furniture and cutlery and other ornamental products, and the pearls that were occasionally found inside the oysters provided an additional source of income. New operations were established as men and boats appeared along the coast. Some, like Tuckey, came from the port of Fremantle in the southern part of Western Australia. Others came from the existing pearling grounds at Shark Bay, or fitted out in Sydney or Singapore before voyaging to the North-west. By the end of 1870 there were thirty boats operating along the coast, employing around sixty whites and between 200 and 300 Aborigines (Bach 1956, pp. 3–9).

At the outset, shell was collected by hand along the shoreline, a practise known as 'beachcombing' or 'dry-shelling'. Tides on the North-west coast are extensive, falling nearly four fathoms (a fathom equals six feet) on a spring tide, exposing extensive coastal flats for several hours each day (McCarthy 2006, pp. 243–59). By the late 1860s, the shoreline around Nikol Bay and the nearby port of Cossack were becoming denuded of shell and beachcombing gave way to free diving, or naked diving, in deeper waters offshore and around the islands that lay off the coast: the islands of the Dampier Archipelago out from Nikol Bay, the chain of small islands that lined the coast between Nikol Bay and the North West Cape; and further afield, at Barrow Island, the largest of the North-west islands, and the islands of the Montebello chain. Pearlshell was now harvested in depths up to six fathoms, the limit that a human was able to descend to, before searching for the Mother-of-pearl oyster and ascending back to the surface (Streeter 1886, pp. 144–63).

This shift to free diving lay at the core of the industry's increasingly bleak character. The standard mode of production involved teams of non-white divers working from small dinghies under the supervision of a white overseer, who, in turn, was attached to a larger ocean-going cutter or schooner. Non-white divers comprised both Aborigines and 'Malays', the name given to South-east Asians brought to the North-west from ports such as Singapore, Batavia, Makassar and Koepang. Aboriginal labour was preferred, not only because of their reputed keen eyesight and stamina, but also because they could be coaxed into diving in return for tea, flour, sugar and tobacco; by contrast, Malays had to be paid in cash and returned to their home islands at the end of a season (Bain 1982, pp. 26–37). But the working life of a diver was often short and, as the industry expanded, it became increasingly difficult to recruit indigenous divers. Many factors contributed to the short working life of a diver. Drowning through accidents or exhaustion or the mistreatment of overseers was common and attacks by sharks were not unknown. Deafness caused by the rigours of repetitive diving debilitated many divers by the end of a season. Cyclones bore down upon the North-west from the northern tropics during the summer season, when diving was preferred owing to the warmer water temperature, periodically wreaking havoc on the pearling fleet. On the water, crowded and unsanitary living conditions and poor diets contributed to the spread of infectious diseases, particularly respiratory diseases such as influenza and measles (De La Rue 1978, pp. 71–82; Withnell Taylor 2002, pp. 114–9). Ashore, a smallpox epidemic, thought to have originated further north in the Kimberley region where Aboriginals were in contact with Makassan trepang fishers, devastated the Ngarla and Ngarluma tribes of the Pilbara coast in 1866, before reappearing amongst pearl divers in 1869–70 and impacting on previously untouched coastal populations (Campbell 2002, pp. 198–214). 'Hundreds of Aborigines died', recorded one observer. 'Their bodies could be found in the mangroves and throughout the country for many months' (Withnell Taylor 2002, pp. 115–6).

During the early 1870s the number of Malays employed increased rapidly. By 1875, when 22 vessels left Koepang and Makassar directly for the North-west coast, at least 989 were known be employed altogether, although the true figure was thought to be as high as 1,800; the number of Aborigines officially engaged in 1875–76 was 493. But the high mortality rates through sickness and accidents and the numerous reports of abuses by white pearling masters that emerged from this season alarmed Dutch colonial authorities in Batavia, leading to a tightening of laws governing the employment of Malays in Australian waters. Their numbers fell away sharply after 1875, increasing the industry's reliance upon the indigenous population. By the 1883–84 there were only 72 Malays, plus another nineteen South Sea Islanders and eleven Chinese working in the North-west. The number of Aboriginal divers that season was estimated at 556 (De La Rue 1978, pp. 77–9).

It was against this backdrop that blackbirding flourished. Slavery had been abolished in the British Empire in 1833 and the Western Australian Governor sought to regulate the employment of Aborigines through statutes passed in 1871 and 1873 but, when cases of mistreatment, kidnap or murder by whites against blacks were brought before the courts, it invariably proved impossible to obtain sufficient evidence to secure prosecution (Forrest 1996, pp. 109–150). This was because most indigenous people could not yet speak sufficient English to understand court proceedings, and because the pearling grounds lay over a thousand miles north of the colonial capital amongst islands that were uninhabited and rarely visited by authorities. Moreover, in the early days of the industry, the legal status of the islands was uncertain. Western Australia's colonial boundaries extended only a league (or three nautical miles) offshore, so most of the North-west islands were in theory unclaimed and beyond the reach of the law. This changed in

1876 when the Governor annexed all offshore islands after a dispute over the ownership of guano deposits on the Lacapede islands that lay off the West Kimberley coastline (Powell 2010, p. 169). By this stage, however, the practise of using islands as baracoons had become established in the North-west. In some cases, islands were used for organised slave markets and in other cases as depots where divers were marooned at the end of the season in the expectation they could be recruited again at the start of the next season. As early as 1871 Faure Island in Shark Bay was reported to be used to detain Aboriginal labour and in 1873 rumours surfaced that a market for female Aboriginals was operating at Barrow Island. More reports emerged during later years of baracoons operating at Barrow Island, Enderby and Delambre Islands in the Dampier Archipelago and the Lacepede Islands (Forrest 1996, pp. 109–110).

As this grim trade continued, raiding parties ashore pushed far beyond the boundaries of pastoral settlements. During the beachcombing phase, labour had been supplied by Aboriginal workers in the slack times between shearing and lambing on the stations and during the transition to free diving much of the labour also came from coastal tribes near Cossack and the other pearling ports of Port Hedland and Onslow. As these populations dwindled through epidemics and the rigours of diving, replacement divers were sought in the newly-opened west Kimberley district or in the remote parts of the Pilbara, Ashburton and Gascoyne districts, where only explorers had ventured previously. Even tribes in the inland ranges on the edge of the vast deserts of inland Australia were targeted and, brought in chains to the coast, were compelled to dive in an ocean they had never before seen (Forrest 1996, pp. 165–8, 192–3). One of the more notorious blackbirders, Thomas Mountain, concentrated his efforts in the headwaters of the Ashburton and Gascoyne rivers and in the area around the North West Cape. Around 1880 he is reported to have made a raid upon the Thalanyi, neighbouring tribe to the Yinikutira (Forrest 1996, pp. 271–7). The traffic of human life to the coast and its islands probably reached its nadir around this time. During 1880 the Governor of Western Australia, appealing to the Colonial Office for approval to introduce stringent new regulations for the industry, stated that the employment of Aborigines in the North-west was 'in a condition little better than slavery' (Bach 1956, pp. 82–3). Mountain was eventually found guilty of illegally detaining Aborigines in 1881, and sentenced to gaol in Fremantle; he never returned to the North-west.

Free diving came to an end during the mid-1880s. Diving suits had been trialled by pearling masters as early as 1868, but did not take off due to the difficulty of diving in a cumbersome 'hard-hat' or helmet in waters marked by large tides and swift currents. However, diving suits were reintroduced to the North-west in 1884 by pearling vessels from Torres Strait, between Australia and New Guinea, where techniques for employing diving apparatus from drifting vessels had been mastered. The take-up thereafter was rapid. There were nine vessels equipped with diving gear in 1884–85, 34 in 1885–86 and 118 in 1887–88, when only two vessels worked exclusively by free diving (Bach 1956, pp. 28–33). The technology – copper helmets, canvas suits, rubber air-hoses and hand-driven air pumps – revolutionised the industry. Divers could now descend deeper, to twelve fathoms and beyond, opening vast new grounds where the Mother-of-pearl shell lay untouched. The pearling fleet moved northwards to the Kimberley coast and the port of Broome, founded in 1883, and adjacent to the deepwater banks that stretched out from Roebuck Bay. A new type of vessel came into use; a double-masted, deep-draught ketch, or 'lugger', around ten to twelve tons and 35 feet in length, suitable for drifting slowly across the banks with a team of divers working underneath. The pattern of labour employed in the industry also changed completely. Aborigines did not take to the diving apparatus, often reacting in horror to the sight of a suited diver, and unable to adapt to the use of the new technology. Skilled Asian divers came to predominate. By 1901 the industry employed 271 Japanese, 705 Malays, 382 'Manilamen' or Filipinos and only 51 Aborigines (Bach 1956, p. 85).

As pearling began shifting to the deeper waters off Broome, the frontiers of settlement in the North-west had been closing. Most of the remaining land suitable for pastoralism, including the North West Cape, was settled during the 1880s and 1890s. The first regular steam service from Fremantle to Singapore via the North-west coast was established in 1885 and by 1889 a telegraph linked the coastal settlements to the outside world (Powell 2010, pp. 171–6). What became of the Aboriginal divers abandoned by the pearling industry? Traditional lifestyle survived only in a restricted sense, and increasingly the Aboriginal population settled in camps and other improvised settlements on the outskirts of pastoral stations, the coastal towns and the mining camps that had emerged in the Pilbara district in the 1880s. Pastoralism provided the main source of employment. Some beachcombing continued and other industries such as a turtle fishery emerged, but the links to a maritime lifestyle that had once marked the introduction to western civilisation for so many of the region's indigenous inhabitants faded quickly after 1885.

Offshore the islands slipped back into a state of isolation. Among the only visitors after free-diving ended and the pearling fleets moved north were naturalists attracted by the remnant populations of marsupials and birds that were becoming increasingly rare, if not extinct, on the mainland. Native fauna had been decimated by the introduction of predators such as foxes and cats, hunting for food and the fur trade and the destruction of habitats by agricultural and pastoral expansion. Disease was also a factor in this decline. An epizootic, thought to have been introduced from South-east Asia by a pearling vessel around Shark Bay in the mid-1870s, had spread across the North-west, interior and South-west regions over ensuing years, devastating populations of the marsupial species held to be characteristic of the Western Australian fauna (Abbott 2006, pp. 530–61). In the late 1890s J.T. Tunney, a collector for the newly-established Western Australian Museum, and widely travelled throughout the colony, located on Barrow Island and two Shark Bay islands, Bernier and Dorre, several species of marsupial that could no longer be found on the continent itself. The zoologist G.C. Shortridge, collecting for the British Museum in 1906, also found on Bernier and Dorre Islands species that were unobtainable elsewhere, and widely believed to be extinct on the mainland. These discoveries led to a movement for the creation of island reserves to protect remnant wildlife populations. An island reserve for native fauna was created at Barrow Island in 1908 (Christensen 2012, pp. 13–30).

Prior to this, in 1907, Barrow Island had been set aside by the Western Australian government as the site for a proposed hospital to treat Aboriginals suffering from venereal disease, tuberculosis and leprosy. Venereal disease had become widespread in the North-west, a legacy of the impoverishment of indigenous society, the spread of prostitution and the breakdown of traditional laws in the context of the fringe camps that now existed across the region. 'Syphilis', although frequently a misdiagnosis of an indigenous form of Granuloma, had in particular reached endemic proportions by the early 1900s (Jebb 1984, pp. 68–86). Leprosy, endemic in South-east Asia, had by this stage reached the Kimberley through contact with Asian pearl divers and the North-west through contact with Chinese miners (Davidson 1978, pp. 12–14). The 'lock hospital', as facilities for the treatment of venereal disease were known, was built during 1908, not on Barrow Island, where aboriginal divers infected with measles had been quarantined in 1883–84, but on Dorre and Bernier Islands, where separate hospitals for males and females were constructed on the twin islands. This left Barrow Island free to become a nature reserve later the same year. A primitive leprosarium was established on Bezout Island, near Cossack, in 1909 (Christensen 2012, pp. 17–22).

The lock hospitals remained open for ten years, often resembling in their operation the earlier traffic of Aboriginal labour to the pearling grounds. Expeditions led by police were sent into far into the Gascoyne, Ashburton and Pilbara districts to locate sufferers of disease, who were then placed in chains and transported across the sea to the remote islands in Shark Bay. The rationale behind this practise combined a rudimentary form of welfare for Aboriginals, who were widely believed to be doomed to extinction, with a fear that venereal diseases could be reintroduced to the white population in a more virulent form. At least 426 patients were sent to the islands, taken from across the North-west and west Kimberley, and at least 162 died there (Jebb 1984, p. 81). In 1918 the hospitals were transferred to Port Hedland on the mainland. Bezout Island continued to receive lepers until 1914, and when the last died in 1917 the islands ceased to be a leper colony, with new leprosariums opening at Port Hedland and Derby in the Kimberley region. Bernier and Dorre Island, protected as temporary fauna reserves during the operation of the lock hospitals, were upgraded to the same status as Barrow Island in 1918, becoming reserves for the protection of flora and fauna.

Fernand Braudel has shown that isolation is a relative phenomenon – the sea is an effective barrier, but island worlds brought within the range of seafarers can be rapidly subjected to historical forces operating on a wider geographical and temporal plane, only to fall away as archaic examples of bygone era if links across the sea are broken (Braudel 1972, pp. 148–51). After 1918 the North-west islands remained uninhabited and largely cut off from the mainland. In 1952, the Montebello Islands were deemed sufficiently remote to host the first British atomic bomb tests. Barrow Island became the site of a major oil and gas development in the 1960s, and again in the 2000s, although the nature reserve remains intact. In 1991 Bernier and Dorre Islands became part of the Shark Bay World Heritage Area and, rarely visited by humans, they remain today as two of Australia's great wilderness reserves. The relics of the Lock Hospitals, like those of the barracoons, remain as evidence of the significance of islands during the conquest of vast North-west (Patterson 2006, pp. 99–111).

The free diving phase of the pearling industry linked these islands to the mainland through the banks of pearshell that lay upon the seafloor and the fleet that sailed these waters to harvest it. Bathymetry, or the distance between the seafloor and the surface, shaped the sea's role as a destructive element in the advance of imperialism in this region. Diving at depth was a dangerous activity, leading frequently to death through drowning, sickness and disease and storms, and driving the industry's demand for labour that was obtained, invariably by force, at increasing distance from the coast. Ultimately it was a technological advance that, by greatly extending this bathymetric limit, pushed the industry into new and deeper waters, where skilled Asian divers took the place of the Aboriginal labour upon which the industry once depended. By then, the frontier of settlement was beginning to close. For the Yinikutira, like other Aboriginal tribes in the North-west, this advancing frontier, presaging as it did the unmaking of their civilisation, was experienced as a three-dimensional space: extending as far inland as the blackbirders raided and as far along the coast as the pearling banks extended, to the islands where labour was bought and sold by the pearling masters and, below the sea, to the depth that a human could be compelled to dive to.

BIBLIOGRAPHY

Abbott, I. 2006. 'Mammalian faunal collapse in Western Australia, 1875–1925: the hypothesised role of epizootic disease and a conceptual model of its origin, introduction, transmission and spread'. *Australian Zoologist* 33(4): 530–61.

Bach, J.P.S. 1956. *The Pearling Industry in Australia: An Account of its Social and Economic Development.* Canberra: Department of Commerce and Agriculture, Commonwealth of Australia.

Bain, M.A. 1982. *Full Fathom Five.* Perth: Artlook Books.

Battye, J.S. 1915. *The History of the North West of Australia.* Perth: V.K. Jones and Co.

Braudel, F. 1972. *The Mediterranean and the Mediterranean World in the Age of Philip II.* Vol.1. London: Collins.

Campbell, J. 2002. *Invisible Invaders: Smallpox and Other Diseases in Aboriginal Australia.* Melbourne: Melbourne University Press.

Christensen, J. 2012. 'Islands as Arks: Nature Protection and the Preservation Ethic', in A. Gaynor and J. Davis (eds) *Ecological Exchanges, Studies in Western Australian History* **29**: 13–30.

Crosby, A.W. 1986. *Ecological Imperialism: The Biological Expansion of Europe.* Cambridge: Cambridge University Press.

Davidson, W.S. 1978. *Havens of Refuge: A History of Leprosy in Western Australia*, Perth: University of Western Australia Press for the Public Health Department.

De La Rue, K. 1978. *Pearl Shell and Pastures: the Story of Cossack and Roebourne and their Place in the History of the North West, from the Earliest Explorations to 1910.* Roebourne: The Cossack Project Committee.

Dench, A. 2009. 'The Aboriginal People of the North West Cape', in S. Skurla, *The Stefano Castaways.* Trans. A. Sala, ed. J. Melville-Jones (Perth: Warrigal Press) pp. 139–50.

Forrest, K. 1996. *The Challenge and the Chance: the Colonisation and Settlement of North-West Australia 1861-1918.* Perth: Hesperian Press.

Hasluck, P. 1942. *Black Australians: A Survey of Native Policy in Western Australia*, 1829–1897. Melbourne: Melbourne University Press.

Jebb, M.A. 1984. 'The Lock Hospitals Experiment: Europeans, Aborigines and Venereal Disease', in Bob Reece and Tom Stannage (eds) European-Aboriginal Relations in Western Australian History. *Studies in Western Australian History* **8**: 68–86.

McCarthy, M. 2006. 'Naked Diving for mother-of-pearl'. *Early Days* **13**(2): 243–59.

Patterson, A. 2006. 'Towards a historical archaeology of Western Australia's Northwest'. *Australasian Historical Archaeology* **24**: 99–111.

Powell, A. 2010. *Northern Voyagers: Australian Monsoon Coast in Maritime History.* Melbourne: Australian Scholarly Publishing.

Skurla, S. 2009. *The Stefano Castaways.* Trans. A. Sala, ed. J. Melville-Jones. Perth: Warrigal Press.

Streeter, E.W. 1886. *Pearls and Pearling Life.* London: George Bell and Sons.

Withnell Taylor, N.E. 2002. *Yeera-muk-a-doo: a Social History of the Settlement of North-west Australia told through the Withnell and Hancock families, 1861 to 1890.* Perth: Hesperian Press.

~5~

Old Waves, New Waves: Changing Objects in Physical Oceanography

Stefan Helmreich

Do ocean waves have a history? The question may sound odd: surely waves are simple facts of nature, matters of the substance of the sea. Waves may have diverse manifestations in marine and maritime lore, a variety of effects on economic and political enterprise and a range of meanings for fishers, surfers and swimmers. But as formal and material entities, the standard view might say, they are best known by a science arriving at ever-improving models of oscillation, undulation and movement.

Historians of oceanography have complicated such a view, documenting the changing systems through which scientists and seafarers have known waves. This essay builds on such work to argue that a cultural history of oceanographic apprehensions of waves can add texture to the sea as a subject for environmental history. I suggest that scholars take particular notice of how the history of wave science has been animated by a persistent concern with forecasting – with looking toward oceanic futures, near-term and, increasingly today, long-term.

WAVES, THEN

The first oceanographic accounts of waves arrive from cultural imperatives in Europe and the United States to know the sea as a site for the extension of land-based activities: shipping, trade, colonial travel, warfare. In the late 1800s, two primary approaches to waves emerged: observational and theoretical. The observational – watching waves – had a centuries-long history in maritime practice. The theoretical began with the treatment of waves as objects for study in hydrodynamic theory. Sir Horace Lamb's 1879 *Hydrodynamics* was a key text, one to which scientists turned as they sought to model waves as perturbations in a 'perfect fluid' (Irvine 2002).

As oceanographic wave science was elaborated in the early twentieth century, researchers sought to fuse the observational and theoretical. In her history of oceanography, *The Edge of an Unfamiliar World*, Susan Schlee (1973) discusses the work of British oceanographer, Vaughn Cornish. In his 1934 *Ocean Waves and Kindred Geophysical Phenomena,* Cornish sought to anchor the theoretical in the observational by calculating – from shore and sea – such measures as wave height and period.

That work of measurement continued in the work of American ocean engineer Willard Bascom, known for creating the wave-energy equation: $Energy = wLH^2/8$. One of the early questions before Bascom, who began his work at Berkeley and Scripps in the 1940s was, 'Did large waves in the ocean do the same things as those in a model tank?' Bascom was keenly aware of – and fascinated by – the difference between model and reality as well as by the sometime dissonance between explanation and experience. In his 1988 memoir, *The Crest of the Wave*, he wrote of his first encounter with the sea, in Humboldt Bay:

> The surf roared at us, which is to say that the wide spectrum of frequencies created by all the waves crashing, colliding, swashing, and releasing bubbles produced a high volume of white noise – a hiss of astonishing proportions broken only by the occasional crack of a large breaker (Bascom, as excerpted in Raban 1993, p. 463).

'Observation', then, entailed not just the visual, but also the sonic. And the olfactory: Bascom was struck by the 'distinctive smell' of the sea. Like many early wave researchers, he revelled in waves as physical things with which it was necessary to wrangle in order to do science. In *The Crest of the Wave*, Bascom narrates a particularly visceral moment, when he is out in an amphibious truck (known nowadays to American tourists as a 'duck boat'), about to be swamped by a wave:

> While balancing under this incipient waterfall, I would estimate the height of the wave that was about to come crashing down, add one third of that … to the trough depth, call the answer into a microphone, and duck. Then the reaching crest of the plunging wave would collapse on us, not quite capsizing the dukw (Bascom, as excerpted in Raban 1993, p. 467; see Figure 1).

A dynamic sense of the entanglement of observation, audition, olfaction, and theory was a signature of Bascom's rhetoric. He wrote in his 1964 *Waves and Beaches: The Dynamics of the Ocean Surface*,

Figure 1. Bascom standing on duck boat with wave ruler, 1940s. Willard Newell Bascom Papers, 1945–2000, Collection 2008–21, BOX 3, Scripps Institution of Oceanography Archives, University of California, San Diego.

> now, full of confidence that we understand waves both in theory and by actual test, we fling open the laboratory door, stride to the edge of the cliff and look to the sea. Good grief! The real waves look nothing like the neat ones … that march across the blackboard in orderly equations. These waves are disheveled, irregular, and moving in many directions. Should we slink back indoors to our reliable equations and brood over the inconsistencies of nature? Never! Instead, we must become outdoor wave researchers. It means being wet, salty, cold – and confused (quoted in Hawk 2005, p. 20).

This is an expression of scientific affect simultaneously at home at the blackboard and outdoors. Bascom's biography can be situated in a history that has seen American natural science researchers seeking to conduct their work in rugged land- and seascapes of the sort institutionalised in the famous work-vacation sites that eventually became Woods Hole, Cold Spring Harbor, Los Alamos and Monterey Bay (see Kohler 2002, Pauly 2000). Of course,

Bascom's declaration of salty confusion may also be a retrospective romantic embellishment, since wave science, particularly by the 1940s in the United States, had become a highly practical endeavour, primarily because of its connection to the war effort.

World War Two brought waves into relief as strategic entities, as oceanographers sought to predict wave action to smooth the amphibious landings of allied troops in Europe and Asia. In 1942, Scripps scientists Harald Sverdrup and Walter Munk were asked by the Oceanographic Section of the Army Air Force Weather Directorate to work on wave forecasting. Their report, *Wind, Waves and Swell: A Basic Theory for Forecasting*, published in 1943, 'found that the size of waves did not depend solely on wind speed but was also affected by the size and shape of the body of water over which the wind blew (the fetch) and the length of time that the wind had been blowing (the duration)' (Schlee 1973, 306; see also von Storch and Hasselmann 2010).

Schlee describes the network of scientists and observers put together in the wake of Sverdrup and Munk's report, a network specially assembled to enable the invasion of Normandy in 1944 (and relocated soon after to Ceylon, to enable the invasion of Burma and Indonesia, which expelled Japan). The 'Swell Forecast Section' comprised 51 wave-reporting stations, at each of which measures were taken of wind speed and direction, wave heights, and periods. A kindred British effort at wave prediction emerged at the Admiralty Research Laboratory, organised by a group of scientists who called themselves the 'W' group (Draper 1996). Predictions for Normandy turned out to be fairly accurate, though currents not yet fully known as well as rough weather complicated actual seashore conditions.

A scientist like Sverdrup might not have been surprised. In *The Oceans: Their Physics, Chemistry, and General Biology*, he had written with colleagues Johnson and Fleming that

> In physics the general picture of surface waves is that of sequences of rhythmic rise and fall which appear to progress along the surface … The actual appearance of the sea surface of the open sea, however, is mostly in the sharpest contrast to that of rhythmic regularity … from the point of view of physics [these real seas] can be termed 'waves' only by stretching the definition (quoted in Irvine 2002, p. 378).

After World War Two, wave science continued to be useful to large-scale institutional activities. Military research continued with researches into wave behaviour during the detonation of nuclear weapons in the south Pacific (see Rainger 2004). The maintenance of oil drilling platforms became important (see Draper 1996 on 1970s oil- and gas licensing enabled by maps of extreme wave heights generated by the UK's National Institute of Oceanography).

Wave science came into its maturity because of human-scaled social and political economic imperatives, imperatives attached to military and commercial forecasting concerns of maritime nation-states. What Philip Steinberg (2001) has analysed as a European model of the ocean as a 'great void' or blank space between nations continued with wave science – as it had begun with wind charts in the nineteenth century – to acquire a surface texture. Wave science would increasingly make visible disturbances on the smooth surface of the blank map – 'When a breeze blows across a stretch of sea, the resulting friction stirs up tiny triangular fluctuations on the water's surface known to sailors as cat's paws and to scientists as capillary waves' (Hawk 2005, 13) – disturbances that could scale up to have large effects, oceanographic and geopolitical both.

A turning point in wave science came in 1961, at the Conference on Ocean Wave Spectra, convened in Easton, Maryland. The major innovation to come out of this gathering of seventy physical oceanographers was the mathematical formalism of the 'wave spectrum'. In a wave spectrum, 'wave records were represented as a weighted sum of sine waves; the relative weightings constituted the spectrum' (Irvine 2002, p. 379). Instead of wiggly lines representing the rise and fall of water, oceanographers turned to a visualisation that broke down a large wave into the smaller, diversely sized waves of which it might be composed. Waves came to be known not as individuals, but as collections of superimposed waves, little and big, with different origins and histories. A 'wave' might be made up of energy generated on some faraway shore, by a hurricane a week ago and by fresh energy from teeny wind-swept ripples. In the 'wave spectrum' model, waves would be rendered not as side views of undulating water, but as collisions of bell curves. The 'spectrum' would be a chart lining up, from left to right, from largest to smallest, the wave sizes (measured in wavelengths) of which a wave was constituted (see Figure 2).

Figure 2 offers a kind of a wavy graph, to be sure, but it is *not* to be seen as a side view of a wave! It is a second-order abstraction. With the arrival of this 1960s formalism, oceanographers stopped talking about individual waves, or about successions of waves, and more about 'wave systems', pro-

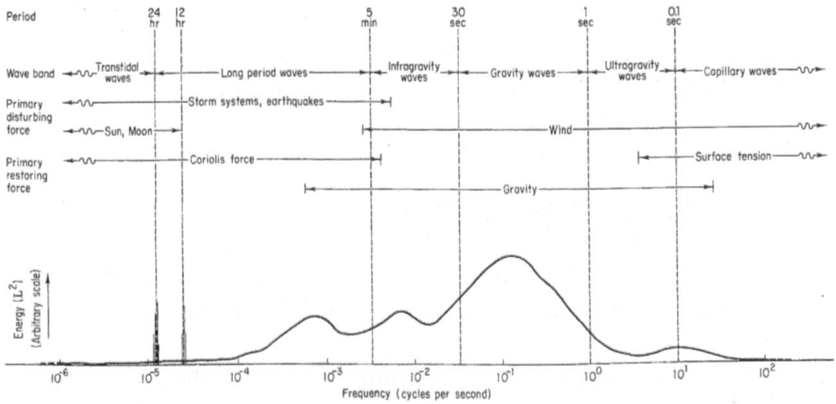

Figure 2. Qualitative wave power spectrum, from Kinsman 1965, p. 23.

cesses that might be understood statistically. The 'nature' of the wavy sea was becoming a question not simply of surveying, navigating and so forth, but a question for the sciences of probability. Insofar as the romantic sentiments of such figures as Bascom were still in the mix, the oceanic sublime was now meeting, and being translated into, the mathematical sublime.

One question for people at the Ocean Wave Spectra conference was: what was the best model for power spectra? And what *produced* spectra? Tides, gravity, wind: all these played a role. But which force was most important? And what were the boundaries of 'waves', anyway? For most of the waves that scientists cared about, wind action inaugurated waves, but should wind be considered part of the ongoing system?[1]

And was there such a thing as a 'typical' wave spectrum? Some conferees assumed that a 'typical' spectrum could be derived from thinking about a relatively smooth and extensive 'fetch' (the sea space across which wind blew at a particular speed) across a bounded 'duration' (the time over which a wind steadily blew).

1 'As Helmholtz reminds us, "The boundary between two fluids of different density cannot remain in equilibrium when they travel at different velocities"' (Lane 1947, p. 63). Whether and how wind stays in the system shapes how observation and theory might be compared (scientists at the Easton conference differed. Some held that waves could travel faster than the wind that generated them; others disagreed). Scientists call 'swells' those bulges of water that travel after initiating wind energy has dissipated.

Others, such as Walter Munk, pointed out that 'fetch' and 'duration' were not things in the world, but formalisations:

> Inasmuch as these terms – 'fetches', 'finite durations' – are really great idealizations of the wind field over the sea, to try and write spectra for given fetches and finite durations is to endow these meteorological notions with more claim to reality than they deserve (quoted in Irvine 2002, p. 380).

Munk's comment points to a key issue in wave science: while waves have a manifest materiality to them, the 'wave' is also an abstraction, one that takes a particular form depending on how waves are conceived, observed, modelled. David Irvine suggests that the character of that abstraction moved into a highly mathematical key with wave spectra; observation – to say nothing of listening, smelling, tasting and swimming! – was sidelined.[2] And when 'observation' *was* discussed, it was already an abstraction. As Irvine puts it, '"Observed" spectra are not really observed; they are the finished products of a sophisticated mathematical analysis ... Spectra stood midway between raw observation and fundamental theory' (2002, p. 382). In this way, spectra share something in common with computer simulations – theoretically animated models within which scientists perform kinds of virtual experimental operations (see Galison 1997). By the middle twentieth century, waves had become scientific objects that could be known through what Daston and Galison (2007) have called 'mechanical objectivity', a representational idiom putatively free of human judgment (even if, for waves, such representations always attached to concerns sculpted by imperatives for prediction).

WAVES, NOW

If power spectra brought wave science into the realm of the mathematised sciences, the process has intensified as wave science has come to make central use of computer simulations and a host of other technical tools. *New York Times* journalist Susan Casey reports that today, 'the study of waves involve[s]

2 The tradition of Melanesian navigation offers a comparison. Anthropologists have suggested that while Melanesian stick and shell charts have been used on land to plan voyages, actual journeying has depended on feeling the sea in which a canoe travels. Taste may also be significant. David Lewis in his 1994 *We, The Navigators: The Ancient Art of Landfinding in the Pacific*, reports on one navigator: 'Kaho is said to have dipped his hand into the sea, tasted the spray and bade his son tell him the directions of certain stars. He then averred that the water was Fijian and the waves from the Lau group where they duly arrived the next day' (quoted in Mack 2011, 128).

quantum mechanics, chaos theory, advanced calculus, vortex turbulence equations and atomic physics' (2010, p. 67). Changing modes of representing waves transform what can be known about them. The kinds of waves that scientists are able to study have expanded. 'Freak' or 'rogue' waves, for example, have become newly legible scientific objects. 'Monster' waves of the sort sought out by extreme surfers have become easier to locate. A rising number of large, unexpected waves that some oceanographers argue may be consequent on anthropogenic global warming have also become visible.

I wish to experiment with calling these exceptional entities 'new' waves. Some are new epistemologically: newly discovered or characterised. Some may be new ontologically: new things in the world.

Let me begin with *rogue waves*. Rogue waves – waves that are undisciplined, abnormal, dangerous, out of place – are waves larger than one might expect from the sea around them. In the statistical idiom of wave sciences, these are waves that leap out of the realm of predictability; a 'wave power spectrum' tailored to explaining wave motion in a particular patch of sea will not forecast them. Such waves have been the stuff of legend, rumour and myth (and in that sense are not new discursively). Ferdinand Lane, in his 1947 *The Mysterious Sea*, writes that, 'Scientists, critical of untrained observers, discount such tales [of larger-than-expected waves], since gravity, they remind us, works steadily to prevent waves from vaulting above a certain height' (p. 65). But in 1995, the first such wave was documented in the North Sea, from a Norwegian offshore monitoring facility for an undersea gas pipeline. In a sea of 39-foot waves, this wave was 84 feet tall (see Parker 2010, p. 128).

New observational practices have made it possible to confirm that such waves exist. And new kinds of models have made it possible to guess at their possible origins. Rogue waves may emerge from the meeting of currents of opposite direction. Off the south-east coast of South Africa, for example, waves from the Agulhas current, which sweeps down East Africa, run into waves from South Atlantic storms coming in just south of Cape Town. '[W]aves forming out at sea off the Cape of Storms can grow almost 100ft tall from trough to crest, about the same as a ten-story building' (Pretor-Pinney 2010, p. 148; and see Rosenthal and Lehner 2008).

Rogue waves may also arise from non-linear effects – from, for example, wave-wave interaction, which can include the transfer of energy from one wave to another. To understand such phenomena, scientists have turned to such formulae as the nonlinear Schrödinger equation, once the province only of scientists working in quantum mechanics. Wave scientists have become ever more dedicated to forecasting using mathematical approaches. To get a flavour of the complexity of this enterprise, consider the following passage from a paper on the European Centre for Medium range Weather Forecasts (which gathers data from buoy systems around the world and feeds data into a wave-modelling programme):

> It appears that freak waves can only arise when the waves are sufficiently coherent. In these circumstances, when the waves are sufficiently steep, nonlinear focusing can act resulting in extreme sea states. These favorable conditions for the formation of freak waves can be quantified by means of a dimensionless parameter called the Benjamin-Feir Index (BFI). Noting the coherency of a wave system can be measured by the width of the corresponding wave spectrum, the BFI is basically a ratio of the steepness of the waves and the width of the spectrum. Large values of the BFI correspond to favourable conditions for freak waves to occur (Bidlot, Janssen, and Abdalla 2006; see also Janssen 2003).

The densely statistical character of this approach links us to a contemporary social domain in which such probabilistic thinking comes to matter: insurance. The global shipping industry operates businesses that embed their financial thinking within the calculus of risk they take to be attendant upon sea travel (see Crane 2010 for a narrative about Lloyds of London's lists of ship losses).

Outsized breaking waves are another sort of wave new to human cultural encounter. Casey describes the 'monster' waves that are increasingly sought out by today's extreme surfers. In the early 2000s, surfers began to look for waves by looking at global satellite information, which they could access quickly (and, now, wirelessly) to make snap decisions about where to go to catch the next wave, accessed through an ecology of standby flights, global-local connections and motor-boats. Bigger waves have not only become newly visible, they have also become newly accessible: 'In recent years, surfers assisted by jet-powered watercraft have been towed into breaking waves … with faces reliably measured in the 60- to 70-foot range. Such swells are too big and fast to paddle into, so the surfers use personal watercraft and water skiing ropes to tow each other into the waves' (Hawk 2005, 90). Monster waves, once the stuff of surfing legend, are now technical and social objects, the stuff of practical expertise (see Figure 3).

Figure 3. 2010 Mavericks competition, by Shalom Jacobovitz – SJ1_8558.

Rogue waves and the waves of extreme surfing are newly discovered, characterised, understood, accessed.[3] Do there exist waves 'new' in a more ontological sense, waves 'new' to the world?

According to Casey's report from the Tenth International Workshop on Wave Hindcasting and Forecasting, held in Hawaii in 2007, many oceanographers predict that climate change will not only change sea level, but sea activity. Climate change means warming water, and warming water is expanding water, which can connect to intensifying wind and storm systems, which can generate systems of higher waves. Not just the substance of the sea is under possible revision by anthropogenic activity (overfishing, acidification, pollution, irradiation), but also, perhaps, its very *form*. Russian oceanographers Sergey Gulev and Vika Grigorieva have suggested that waves may be getting taller. They suggest that waves in the North Atlantic have been rising fourteen centimetres each decade since the 1950s, while

3 Other waves new to science might include 'internal waves,' underwater waves composed of water of one density and temperature meeting another.

the Pacific has seen increases of eight to ten centimetres (Gulev and Grigorieva 2004). A few scientists have begun to ask whether the numbers of rogue waves may be on the rise, with increased frequencies and strengths of storms and with more interaction between storm systems (Bitner-Gregersen and Toffoli 2013).

We can look anew at Helen Rozwadowski's summary of how the sea has been rhetorically positioned outside human history: 'Although most glimpses out to sea reveal endless waves reaching to the horizon rather than any lasting evidence of human presence, there are myriad human activities through which historical connections between people and the oceans can be explored' (2010, 162). In the contemporary world, even 'endless waves' bear the marks of human agency. Human engagements with waves might no longer be imagined as a straightforward encounter of culture with nature. Waves have become hybrids of nature–culture.

Notable about these 'new' waves is their 'extreme' character. A survey of the proceedings of the Eleventh Workshop on Wave Hindcasting and Forecasting (in 2009) sees dozens of papers that use 'extreme' in their titles – 'extreme waves', 'extreme sea states', et cetera. The 'extreme' modifier is not simply empirical; it is a token of recent fascinations with things extreme, part of the same zeitgeist that has recently given us two popular science books with the title *Extreme Nature* (Curtsinger 2005, Cowardine 2008), part of the same episteme that sees marine biologists fascinated with *extremophiles*, organisms that thrive at extremes of temperatures, pressures, salinities and so on (Helmreich 2009). In the current day, the 'extreme' has become a frame for thinking about nature at its boundaries.

I indicated at the opening of this paper that the history of wave science has been bound up with forecasting – and therefore with the *future*. As a figure for environmental history, then, the wave offers a marker of how nature is brought into focus as an unfurling force. The difference between old and new waves, seen in this light, might be the lengths of futures that scientists in different historical periods imagine. For old waves, futures were near-futures (of wartime, weather forecasting). For new waves, these sorts of futures are implicated as well but, with 'extreme waves', there is a longer-term future at issue, too, one to do with century-long predictions about climate and its transformation.

Geographers Jon Anderson and Kimberley Peters note that 'geography' is 'earth-writing' (2014, p. 3). Insofar as oceanography – 'ocean-writing'

– has entered geography, it has not been as a companion interpretative, humanities-styled inquiry, but rather as a science. I hope here to have sketched out the beginnings of an environmental historical oceanography, one that considers waves as historical and cultural objects, subjects for an inquiry we might call 'wave-writing', which, to borrow from the Greek for wave, κύμα, we could name *kymography* – though since that word already exists to describe the tracing of cardiac pulse waves, we might nominate with a more unholy Latin+Greek neologism: *undulography*. Undulography will map out how the history of the sea has prepared a variety of oscillating futures, natural and cultural.

BIBLIOGRAPHY

Anderson, J. and K. Peters. 2014. *Water Worlds: Human Geographies of the Ocean*. Farnham: Ashgate.

Bascom, W.N. 1964. *Waves and Beaches*, excerpted in J. Raban (ed.) 1992. *The Oxford Book of the Sea*. Oxford: Oxford University Press. pp. 468–471.

Bascom, W.N. 1988. *The Crest of the Wave: Adventures in Oceanography*, excerpted in J. Raban (ed.) 1992. *The Oxford Book of the Sea*. Oxford: Oxford University Press. pp. 462–468.

Bidlot, J-R., P. Janssen and S. Abdalla. 2006. 'Extreme Waves in the ECMWF Operational Wave Forecasting System'. 9th International Workshop on Wave Hindcasting & Forecasting, Victoria, B.C., Canada, 24–29 September 2006.

Bitner-Gregersen, E.M. and A. Toffoli. 2013. 'Probability of Occurrence of Rogue Sea States and Consequences for Design'. Proceedings of the 13th International Workshop on Wave Hindcasting and Forecasting and 4th Coastal Hazards Symposium, Banff, Canada, 27 October–1 November 2013: http://www.waveworkshop.org/13thWaves/index.htm

Carwardine, M. 2008. *Extreme Nature*. New York: Harper.

Casey, S. 2010. *The Wave: In Pursuit of the Rogues, Freaks and Giants of the Ocean*. New York: Doubleday.

Cornish, V. 1934. *Ocean Waves and Kindred Geophysical Phenomena*. Cambridge: Cambridge University Press.

Curtsinger, B. 2005. *Extreme Nature: Images from the World's Edge*. White Star.

Daston, L. and P. Galison. 2007. *Objectivity*. New York: Zone Books.

Draper, L. 1996. 'The History of Wave Research at Wormley, A Personal View'. *Ocean Challenge* **6**(2): 24–27.

Galison, P. 1997. *Image and Logic: A Material Culture of Microphysics*. Chicago: University of Chicago Press.

Gulev, S. and V. Grigorieva. 2004. 'Last Century Changes in Ocean Wind Wave Height from Global Visual Wave Data'. *Geophysical Research Letters* **31**: L24302, doi:10.1029/2004GL021040.

Hawk, S. 2005. *Waves*. San Francisco: Chronicle Books.

Helmreich, S. 2009. *Alien Ocean: Anthropological Voyages in Microbial Seas*. Berkeley: University of California Press.

Irvine, D. 2002. 'The Role of Spectra in Ocean Wave Physics', in K.R. Benson and P.F. Rehbock (eds) *Oceanographic History: The Pacific and Beyond*. Seattle, WA: University of Washington Press. pp. 378–386.

Janssen, P. 2003. 'Nonlinear Four-wave Interaction and Freak Waves'. *Journal of Physical Oceanography* **33**: 863–884.

Kinsman, Blair. 1965. *Wind Waves: Their Generation and Propagation on the Ocean Surface*. Englewood Cliffs, New Jersey: Prentice-Hall, Inc.

Kohler, R.E. 2002. *Landscapes and Labscapes: Exploring the Lab-Field Border in Biology*. Berkeley: University of California Press.

Lane, F.C. 1947. *The Mysterious Sea*. New York: Doubleday.

Mack, J. 2011. *The Sea: A Cultural History*. Reaktion Books.

Parker, B. 2010. *The Power of the Sea: Tsunamis, Storm Surges, Rogue Waves, and Our Quest to Predict Disasters*. New York: Palgrave Macmillan.

Pauly, P.J. 2000. *Biologists and the Promise of American Life: From Meriwether Lewis to Alfred Kinsey*. Princeton: Princeton University Press.

Pretor-Pinney, G. 2010. *The Wave Watcher's Companion: From Ocean Waves to Light Waves via Shock Waves, Stadium Waves, and All the Rest of Life's Undulations*. New York: Perigee.

Rainger, R. 2004. '"A Wonderful Oceanographic Tool": The Atomic Bomb, Radioactivity and the Development of American Oceanography', in H.M. Rozwadowski and D.K. van Keuren (eds) *The Machine in Neptune's Garden: Historical Perspectives on Technology and the Marine Environment*. Sagamore Beach, MA: Science History Publications. pp. 96–132.

Rosenthal, W and S. Lehner. 2008. 'Rogue Waves: Results of the MaxWave Project'. *Journal of Offshore Mechanical and Architectural Engineering* **130**(2): 021006, doi:10.1115/1.2918126.

Rozwadowski, H. 2010. 'Playing By – and On and Under – the Sea: The Importance of Play for Knowing the Ocean', in Jeremy Vetter (ed.) *Knowing Global Environments: New Historical Perspectives on the Field Sciences*. New Brunswick, NJ: Rutgers University Press. pp. 162–189.

Schlee, S. 1973. *The Edge of an Unfamiliar World: A History of Oceanography*. New York: Dutton.

Steinberg, P.E. 2001. *The Social Construction of the Ocean*. Cambridge: Cambridge University Press.

von Storch, H. and K. Hasselmann. 2010. *Seventy Years of Exploration in Oceanography: A Prolonged Weekend Discussion with Walter Munk*. Berlin: Springer-Verlag.

Sverdrup, H. and W. Munk. 1943. *Wind, Waves and Swell: A Basic Theory for Forecasting*. La Jolla, CA: Scripps Institution of Oceanography.

~ 7 ~

Fishing Communities in Eighteenth-Century Iceland – An Interdisciplinary Case Study from Vestfirðir

Stuart Morrison

INTRODUCTION

As an island nation, isolated from Europe by the chilly waters of the North Atlantic Ocean, it is easily understandable that the maritime environment played a significant part in shaping Icelandic society. Through the centuries following the *Landnám* (Old Norse: Land-take) of Iceland in the ninth century AD, largely by settlers of Nordic origin, fishing and maritime resource exploitation provided a vital commodity for both subsistence purposes and for trading, initially within the country itself before extending into the developing wider European trading network.

A recent interdisciplinary study examining the development of early fishing communities in the Vestfirðir region, the peninsula located at the north-west extremity of the country, reveals distinct and unique episodes in the emergence of fishing across a number of locations, spanning around eight centuries (Morrison 2012). This study has begun to develop a picture of maritime history in Vestfirðir and has highlighted the variations in type of activity that can be observed within coastal communities in one geographically specific region at different periods (Figure 1).

One of these distinct chronological periods is the eighteenth century. The traditional historical narrative has seen the seventeenth and eighteenth centuries as periods of decline in the Icelandic fisheries (Robinson 2009, Ogilvie and Jónsdóttir 2000). This is somewhat coloured by contemporary perspectives when being discussed against perceptually more economically successful periods such as the late eighteenth and early nineteenth centuries. One scholar who visited Iceland in the eighteenth century complained that the farmers were more interested in fishing than farming, putting minimal effort into agriculture (Edvardsson 2005).

That landed interests were considerably more in favour of agriculture over the maritime economy (Karlsson 2000, p. 154) is reflected in the banning of worm bait lines with multiple hooks, the deliberate stifling of investment for fear of losing control of resources and labour (Vasey 1996) and the imposition of tax; fisheries were taxed while agriculture was exempt. In addition to this last point, agriculture was in receipt of over twenty times more government aid than the fisheries in the following century (Jónsson 1983), a century when opinions had changed and commercial fisheries were fully embraced from an economic perspective.

The aforementioned study involved excavation at five sites in the Vestfirðir region, two of which revealed evidence of earliest occupation around the middle of the eighteenth century. Instantly we see evidence for an episode of apparent development on the coast of Vestfirðir during a century of supposed decline and stagnation. This initial discovery raises the question – was this century really one of decline, or has a reduced episode of regional development been lost in the wider context of national economic decay?

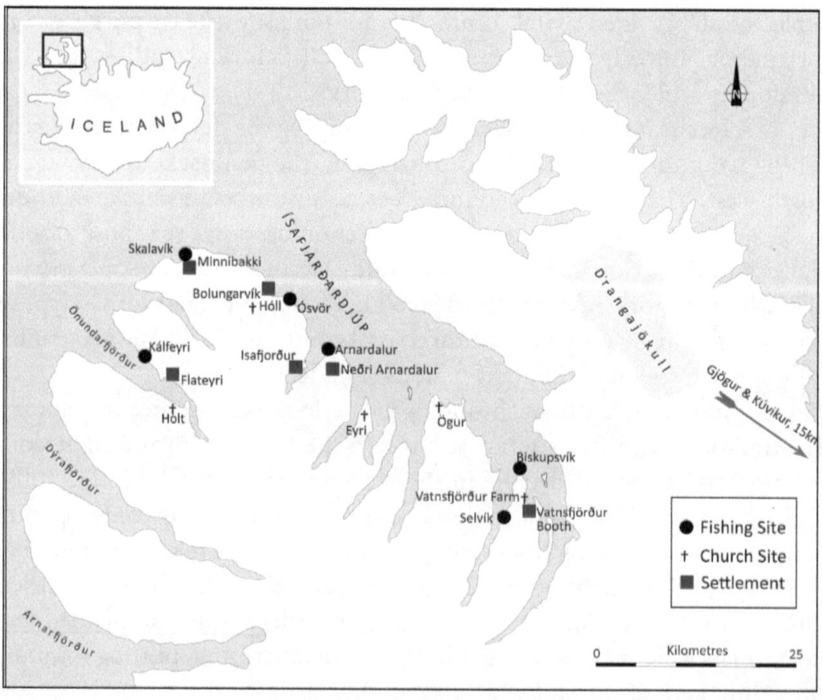

Figure 1. Key study sites within Vestfirðir, Iceland.

THE EIGHTEENTH CENTURY SITES

The excavations and analysis that reveal two sites as emerging at some point around the mid-eighteenth century is more than a simple indicator of presence and absence. Through the application of an integrated geoarchaeological methodology, the cultural record that is retained within the soils and sediments can begin to be systematically and chronologically investigated to build a more detailed narrative.

The first identified site is Biskupsvík (Figure 2), located deep within the Isafjarðarjúp fjord, close to its junction with Vatnsfjörður fjord. A site with the primary purpose of landing fish in this area is mentioned in *Jarðabók,* published in 1712, a land registry compiled for taxation purposes (Magnússon and Vídalín 1940). What was landed here, according to *Jarðabók* was more for subsistence purposes and of *little or no commercial value* (Edvardsson, personal communication). Biskupsvík is ideally positioned for catching a mix of freshwater and saltwater species within the fjord.

Given the range of both size and species that this mix implies, the inconsistency in catch renders it more difficult to preserve and market,

Figure 2. The Biskupsvík site.

hence more suitable for domestic consumption. The geoarchaeological record reveals an earliest calibrated radiocarbon age of AD 1715–85 (Figure 3). Total phosphorus and magnetic susceptibility, two of the strongest indicators of anthropogenic influence in soils and sediments, record high levels within the stratigraphy. Thin section micromorphological analysis

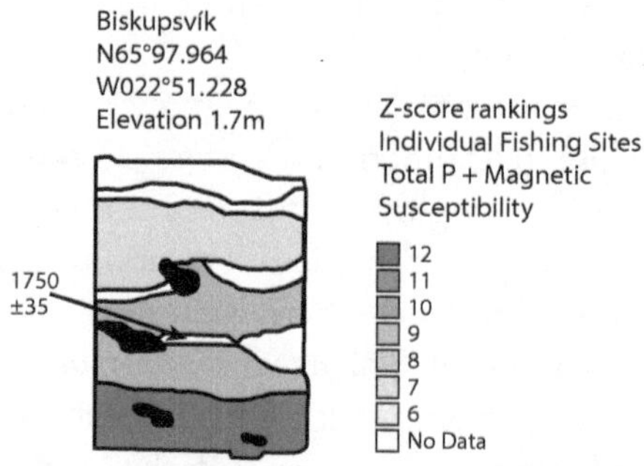

Figure 3. Biskupsvík stratigraphy, detailing radiocarbon dates, total phosphorus and magnetic susceptibility Z-score rankings.

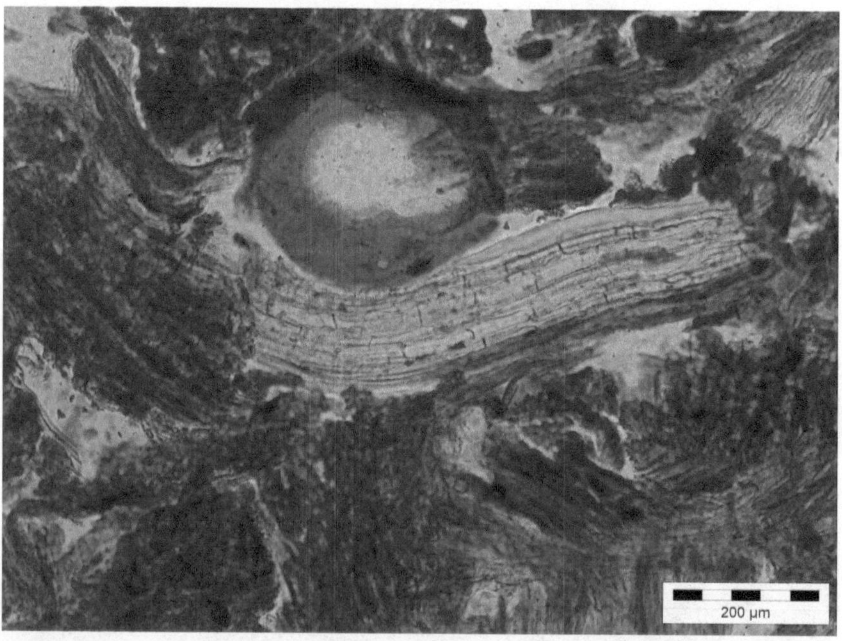

Figure 4. Plant remains found at Biskupsvík. 10x magnification, plane polarised light.

reveals that the site, once occupied, was subject at first to very intense use but that this gradually dwindled over time, as indicated by the volume of bone and plant material preserved within a matrix of domestic (as opposed to industrial) debris (Figure 4).

The second site, Arnardalur (Figure 5), is located on the eastern shore of Skutulsfjörður fjord, close to modern day Isafjörður. Historical sources first mention the site in the eighteenth century (Edvardsson 2010, p. 170). The records reveal that it was a fishing station and the property of the nearby privately owned farm, Neðri-Arnardalur, with nothing paid to the church apart from tithes. Tolls obtained from the fishing station were paid instead to the main farm. Close by is Fremri-Arnardalur, which translates as *foreign exchange in Arnar's Valley*, highlighting the role this site played in the eighteenth century as a connection between Iceland and wider trading centres: Arnardalur was an eighteenth-century commercial operation, firmly within a wider European trading context.

Figure 5. The Arnardalur site.

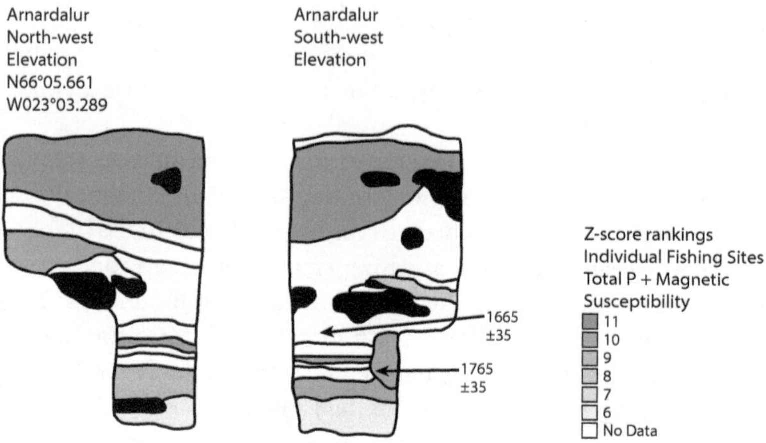

Figure 6. Arnardalur stratigraphy, detailing radiocarbon dates, total phosphorus and magnetic susceptibility Z-score rankings.

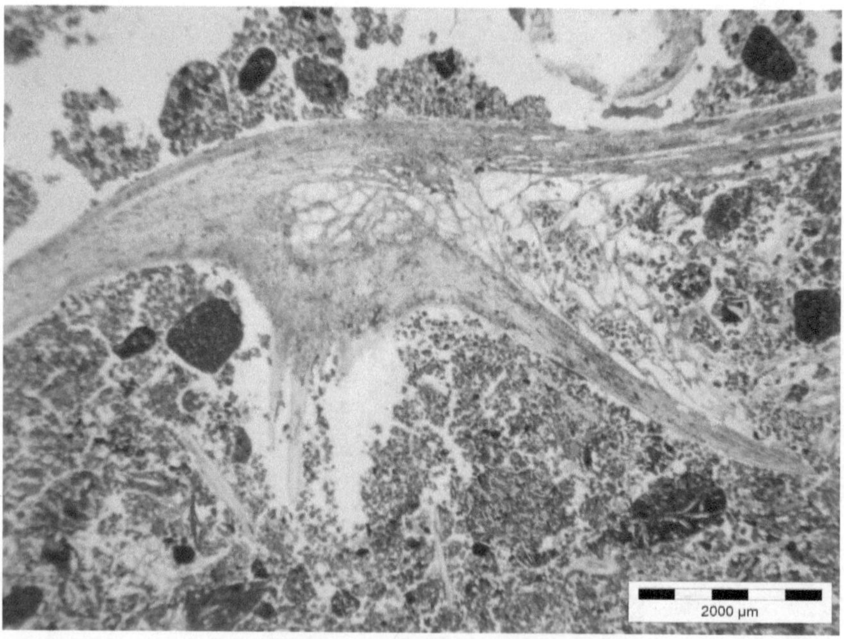

Figure 7. Bone material found at Arnardalur. 1.25x magnification, plane polarised light.

The site itself is suitably positioned on the edge of a peninsula, for easy access to the deepwater fishing grounds and return to shore in the shortest time possible, essential in order to maximise opportunity within brief windows of favourable weather. Fish obtained from these fishing grounds were of a higher economic value than at any other location. The species caught was cod, and of a fairly uniform size, between sixty and 110 centimetres in length. This standard size was not only ideal for preservation, but also more marketable and easier to process due to greater levels of consistency. The geoarchaeological record reveals a site with a lowest calibrated radiocarbon age of AD 1730–1800 (Figure 6), which experienced distinct phases of occupation and abandonment, coupled with a transition towards reduced bone amongst sustained levels of domestic debris (Figure 7), perhaps reflecting technological advancement within fish processing rather than decline in the fishing industry. With processing taking place offshore, a reduction in the maritime signal can be expected within a soil matrix generally containing an increasing proportion of domestic debris. For further explanation of fuel residues and their interpretation, see Simpson et al., 2003.

EXPLANATIONS BEHIND EMERGENCE

The two sites discussed both emerge around the same time, within the same geographical region. They are highly contrasting in terms of position within the fjord system and also in the ways in which they functioned in Icelandic society; Biskupsvík was more for meeting subsistence needs, with Arnardalur meeting commercial economic demand. With both sites active around the same time, but contrasting so significantly in their archaeological content, a wider understanding of the driving factors needs to be assessed. The following section will draw together an outline environmental and social narrative in order to explain their development and to establish the function and significance of these sites and the potential drivers that instigated this transition.

A BRIEF ENVIRONMENTAL AND SOCIAL HISTORY

The persistently inclement climatic conditions and the frequency and magnitude of tectonic activity, coupled with the outbreaks of pestilence amongst humans and livestock, reveal the eighteenth century to be one of the bleakest in Icelanders' experience. It was a century where records were being broken for the wrong reasons and the worst experiences during historical time were being endured. Incidents where natural agency negatively affected the human population were numerous; the eighteenth century was the most volcanically active century since the *Landnám* (Larsen and Eiríksson 2008); the eighteenth and nineteenth centuries saw Iceland experience the most severe climatic conditions, with the extents of sea ice and glaciers reaching their maximum (Axford et al. 2009); smallpox and famine (associated with environmental hardships) created sharp drops in the population, with two of the worst crises to hit Iceland during its history taking place during this period (Jackson 1982, Vasey 1991, Vasey 1996, Ogilvie 1981, Karlsson, 2000). Contemporary writers estimate that, during the century, Iceland endured 43 years of hardships (Stephensen 1806).

The eighteenth century began with the climatically mildest decade of the century (Ogilvie 1986) but, within this, Iceland had to contend with a smallpox epidemic that lasted from 1707–09. Killing around a third of the population, this was the worst population crisis to face Iceland (Jackson 1982, Vasey 1991, Agnarsdóttir 2008). Iceland's first census in 1703 allows for reasonably high levels of numerical accuracy regarding the population from this period onwards (Figure 8) and subsequent censuses reveal a decline of around twelve per cent from the 1703 high. The drop in available labour that this represented had a profound impact on the island's economy.

Year	Population
1703	50,358
1734	43,377
1751	48.799
1755	48,000
1759	42,822
1778	49,863
1783	49,753
1785	39,251
1786	38,363
1791	40,130
1801	47,852

Figure 8. Iceland's Population, 1703–1801 (Data obtained from Ogilvie 1984, Jackson 1982, Vasey 1991, Karlsson 2000).

Iceland was a country whose function was dependent on the availability of a highly mobile labour force. Maritime and terrestrial economies went hand in hand; fishing was undertaken in the winter and spring (in the south and west), with farming duties completed in the summer and autumn across the country. Many men migrated to the south and west to work on the rowing boat fisheries and back to farms when the season was over (Thór 2009). Many within the group of labouring individuals took the opportunity to reside in the farms made vacant by those succumbing to the smallpox epidemic, seizing the opportunity to move up the Icelandic social ladder and, in turn, create a further shortage of available labour. Farmers in coastal regions found it difficult to get people to man their boats (Ogilvie 1984). Had the Danish Crown not passed a law in 1706 forbidding boats within waters deemed to belong to Iceland and the overwintering of Danish merchants (Thór 2009) – in other words, had foreign labour been permitted to move in – the shift from the coastal fisheries to the vacant farms might not have created such widespread coastal abandonment. Decline of the coastal economy, in this instance, appears to have been enhanced by the contribution of unfavourable rulings by the Danish Crown. This was not the first time, as a Danish monopoly over Iceland had been imposed in 1602.

The fisheries during this era were not overly productive, adding to the temptation to develop a farming economy. The colder temperatures witnessed during the eighteenth century meant that cod migrated southwards in order to find waters between four and seven degrees centigrade, which the species prefers (Ogilvie 1981). Sea ice increases further exacerbated this problem, as this increasingly cools the surrounding waters. Given the difficulties witnessed between 1685 and the end of the first decade of the eighteenth century, the high levels of risk in setting up fisheries at a distant location might well have resulted in the same outcome of widespread coastal abandonment through poor fisheries. In addition to this, dried fish was not yielding a high price in Europe during the seventeenth and eighteenth centuries (Thór 2009), particularly when compared to the Golden Era of fisheries between 1300 and 1550. This would have brought a disincentive to invest and, in a way, supports the idea that Iceland benefitted from the monopoly by not being subject to the fluctuations of a distant market.

The concurrent bad years in the 1750s, particularly 1751–57, brought with them enormous levels of human suffering through year on year pressures adversely affecting the terrestrial and maritime landscape (Ogilvie 1991). The arrival in quick succession of inclement environmental conditions allowed for little recovery time. This position was compounded by the impact of regular livestock crises in the country; in the 36 years between 1730–66, livestock died from hunger in 24 (Ogilvie 1984). More cattle were dying year on year. Around the mid 1750s, the fisheries in the west were noted as being particularly poor (Bréf til Stiftamtmanns úr Dalasýslu 1704–1785, in Ogilvie 1981). These episodes would have placed increased pressure on already dwindling resources. The eruption of Mount Katla in 1755 added further misery (Jackson 1982, Vasey 1991). Ice was preventing fishing, notably in 1756 (Jónsson 1994), and harvesting hay on land was simply not possible (Ogilvie 1981), resulting in famine between 1756–58 (Vasey 1991, Oslund 2004). The fisheries decline extended across the entire period from 1744–59 (Ogilvie 1984), with human deaths ever-increasing towards the end of that period. It is estimated that between 1751–59, around 6,000 people died as a result of starvation (Jackson 1982). Problems were Iceland-wide, with people migrating from the east and north to the west and south-west, taking the few remaining resources belonging to the original inhabitants (Ogilvie 1984).

The environmental disaster that faced Iceland was recognised overseas. In 1763, regulations were passed allowing foreigners to enter Icelandic waters (Eggertsson 1996); however they were not allowed to employ Icelanders, possibly for fear of drawing labour away from the already fragile terrestrial-based economy. This limitation reflects the control that the landed elite had over Iceland's economy and their stranglehold over maritime interests and national decision-making. Even with a reprieve from the social and climatic perspectives, the following decade still brought with it environmental challenges; heavy sea ice in 1766 was compounded by the eruption of Hekla, which lasted for two years, resulting in further profoundly negative terrestrial impacts (Ogilvie 1986). This period of difficulty was in some ways timely where development opportunities existed, as good fisheries are widely reported from 1766–79, correlating well with the occupation of Arnardalur. The size of the fishing fleet grew rapidly: by 1770 there were 1869 registered vessels and 9,027 fishermen in Iceland (Thór 2009). Boats, however, got progressively smaller from the 1770s onwards as smaller craft became more favoured. This was a result of a drop in supplies of boats from Denmark and the drop in the coastal population of Iceland. The boats that required fewer men, generally between four and six, addressed the labour shortage but were restricted in the distance that they could travel from shore: twenty kilometres is as far as a boat of this size could realistically travel on a single trip (Edvardsson et al. 2004). This range suited the fishermen of Vestfirðir, given that the more productive fishing grounds are located within a short distance of the heavily fjorded shoreline (Edvardsson et al. 2004, Ogilvie and Jónsdóttir 2000).

High, sustained catches did not always result in commercial success; the processing method used by Icelanders since *Landnám,* the air-drying of fish, relied on favourable climatic conditions in order to complete the process. Conditions in Reykjavík in the 1760s were not so favourable and, despite successful fisheries, the persistent rain resulted in large volumes of cod being spoiled, therefore being unsuitable for export and going to waste (Ogilvie 1984, Jónsson 1982). Instances such as this may have been a catalyst to adopting a new method of preservation – salting. The company in charge of the monopoly of Iceland's trade in 1766, *Almenna verslunarfelagid*, sent a representative to show Icelanders how to salt fish. Merchants were keen to exploit the Spanish and Italian markets of the time (Thór 2009), which were opening up because of changes in supply from the Newfoundland fisheries. Icelanders had no previous experience in the use of salt, due to its high production and importation costs; however the technique spread to such an extent that the majority of fish producers in Iceland were salting by 1790 (Thór 2009). No longer were Icelanders bound by the seasonality of the stockfish 'window', nor dependent on favourable climatic conditions to preserve their catch.

Even with the prolonged period of favourable fisheries, Iceland in the 1770s was said to have lacked resilience and thought to have been close to socio-economic collapse (Eggertsson 1996). The country was in the midst of a twenty-year-long sheep epidemic, which peaked in the 1780s and coincided with a sequence of cold winters. The origins of this epidemic seem to have been in the 1750s at an experimental farm in Reykjavík, where a number of English rams imported for breeding purposes were said to have introduced the disease to the island (Karlsson 2000, p. 177). This event would likely not have taken place had it not been for the investment made in the 1750s to promote textile industries in Iceland (Oslund 2004), another reflection of the terrestrially-focused direction the landed elite wanted Iceland to take.

The 1780s were remarkable in terms of hardship witnessed in Iceland. Through the influence of scientific observers who visited around this period, however, people in mainland Europe were learning more about the tribulations of this distant island. In 1782, the seas were frozen and the homefields 'bald' (through year-on-year biomass offtake with minimal regeneration), severely affecting livestock productivity, exacerbating landscape degradation and preventing fishermen from obtaining any stock. Landscape degradation was accelerating (Morrison 2012), highlighting additional pressures on the grazing areas. Degradation during this period is observed in the oceanic record that shows greater levels of sediment influx into the fjords and seas around Iceland (Jennings et al. 2001). In Ísafjarðarsýsla, hardships continued into the following year, with livestock struggling to obtain pasture (Ogilvie 1984); however the additional environmental catastrophe of tectonic activity was to affect the whole of the country.

In January 1783 (Oslund 2002), a fissure close to Reykjavík began to erupt, but the year is most notable for the Laki eruption in June. Sulphur-rich rainfall destroyed grasslands, fluorine probably destroyed much of the inland fishery and the coastal fisheries failed, not recovering until mid-spring 1784 (Vasey 1991). Some regions of Iceland were affected worse than others, resulting in mass internal migration to avoid those areas worst hit by the *Móðuharðindi*, or 'Mist Hardships' (Karlsson 2000, p. 180). A haze veiled not only Iceland, but also the skies over Western Europe. The impacts were such that it is thought this eruption (alongside the poor climatic conditions witnessed in the years immediately previous) disrupted harvests across Europe, so much so that it is regarded as a potential catalyst for the French Revolution (Karlsson 2000, p. 180). Mortality levels for 1783 do not show a sharp increase, indicating that it was the longer-term impacts of the eruption that caused the greatest effects on the population.

In Ísafjarðarsýsla, November 1784 saw death rates rise sharply through an outbreak of disease, exacerbated by hunger and starvation caused by the hardships (Vasey 1991). Save for this, mortality in Vestfirðir was reported to be lower than in the rest of Iceland (Vasey 1991), possibly because of its increasing dependence on maritime over terrestrial resources, enabling the regional population to respond quicker to the environmental pressures caused by the Laki eruption. It is this shift that correlates best with the emergence of Biskupsvík.

Terrestrial resources were slower to respond; but their destruction was at unprecedented levels. It is estimated that 10,000 cattle, 27,000 horses and 190,000 sheep, representing fifty, 75 and eighty per cent of their respective species populations, were said to have perished (Jackson 1982). This scale of loss created enormous demand for alternative foodstuffs to be obtained, with starvation widely recorded. The livestock that remained, if they could not be supported by the landscape, were slaughtered (Ogilvie 1984), reflecting an attitude conditioned by periodic hardships.

The socio-economic prevalence of tenantry further compounded the impact on the landscape driven by environmental factors. The absence of an ownership mentality provided less of an incentive to utilise the landscape in a long-term and sustainable manner. Tenants were planning for one season at a time and making the most out of what they had right then, rather than considering several years into the future. This was a time of increased difficulty and, with this additional pressure, the population was pushed towards the coasts to meet their resource demands.

BRIEF ECONOMIC OVERVIEW

The Danish monopoly over Iceland's external trade has been viewed as having both positive and negative impacts on Iceland. It was said to have brought stability over the longer term, with guaranteed prices on goods but, with trade and exchange being constant, many felt this hindered economic developments and stifled creativity. Nevertheless, before it was lifted, few actually expressed a desire to have it abolished (Karlsson 2000, p. 142). The sense of past negativity was likely exaggerated by contemporary writers due to the rapid economic growth that was observed once the monopoly was lifted in 1787.

With the country enduring such levels of recorded hardship while also witnessing a boom period through increased economic demand in the second half of the eighteenth century, both internally and from overseas, a stark contrast can be observed at coastal sites. Regulations passed by Danish governors in 1763 allowed foreigners into Icelandic waters for fishing, a decision likely to have stemmed from European pressure caused by the hardship the continent was facing in the 1760s. The Dutch were swift to capitalise on this change of policy, with their activity in Icelandic waters peaking in the 1770s (Robinson 2009). This policy change may have been the economic stimulus that channelled labour efforts to the Icelandic coasts, an effect further enhanced after the arrival of decked vessels in the 1790s, which meant fish could effectively be 'hunted down' and crews could stay at sea for longer. This technological advance was paired with salt curing.

EIGHTEENTH CENTURY ICELAND IN A MODERN GLOBAL CONTEXT

Many currently prevalent themes in society have direct correlations with the experiences of Iceland during the eighteenth century. Evidence reveals a society that was, to a point, resilient but still marginally close to collapse, and also one that was responsive to the hardships it faced but equally stifled by environmental conditions and political decisions.

The comparatively low extraction levels of fish from the seas suggest that overfishing was not an issue. That, however, is not to suggest that the modification of riverine environments through sediment inundation did not have negative impacts on the freshwater fish population. A decline in the freshwater catch increased demand on resources from saltwater environments, placing reliance on a single environment. This trend is observed in the twelfth century, and is said to have stimulated coastal activity then, and may have done so again during this later period (Morrison 2012).

The issues affecting the Icelandic landscape, coupled with the transition of focus from the land to the seas, as reflected in the sites investigated here, may have given Icelanders a stimulated opportunity to explore their coasts. What is alluded to in the documentary record is reflected more clearly in these two coastal sites. With developments in marine technology allowing safer exploration away from land, it may have been the first time since *Landnám* that Icelanders were rediscovering their traditional connection with the ocean. This society, driven by explorers, used the seas as their primary means of navigation. They lost their maritime connection during the Middle Ages, but with the growing use of boats and developing international trade connections, were given the chance to reconnect. Icelanders were, at this time, conscious of the benefits of technological developments. Whereas in their ancestral past they had been at the mercy of their Goðar, largely permitting their beliefs to shape the outcome of their lives, now they were entering a period of science and enlightenment. The landscape and seascape were to be used as man saw fit and to be explored with the only limitations being personal and technological capability. The levels of technological advancement and exploitation of marine environments that took place during the subsequent two centuries support this change in attitude. But now the approach has shifted back again, to viewing the seas as a precious resource to be sustainably managed. In consequence, hundreds of thousands of square kilometres around Iceland represent one of the richest, but equally one of the most protected, seascapes in the world today – the envy of many countries who find their seas depleted and devoid of a rich maritime habitat.

BIBLIOGRAPHY

Agnarsdóttir, A. 2008. 'The Danish Empire: The Special Case of Iceland', in M.N. Harris and C. Lévai (eds) *Europe and its Empires*. Pisa: Pisa University Press. pp. 59–84.

Axford, Y., Á. Geirsdóttir, G.H. Miller and P.G. Langdon. 2009. 'Climate of the Little Ice Age and the past 2000 years in northeast Iceland inferred from chironomids and other lake sediment proxies'. *Journal of Palaeolimnology* **41**: 7–24.

Edvardsson, R. 2005. 'Commercial and Subsistence Fishing in Vestfirðir: A Study in the Role of Fishing in the Icelandic Medieval Economy'. *Archaeologia Islandica* **5**: 51– 67.

— 2010. 'The Role of Marine Resources in the Medieval Economy of Vestfirðir, Iceland'. Unpublished Ph.D. Thesis, City University New York.

Edvardsson, R., S. Perdikaris, T.H. McGovern, N. Zagor and M. Waxman. 2004. 'Coping with hard times in NW Iceland: Zooarchaeology, history and landscape archaeology at Finnbogastaðir in the 18th century'. *Archaeologia Islandica* **5**: 20–46

Eggertsson, T. 1996. 'No experiments, monumental disasters: Why it took a thousand years to develop a specialized fishing industry in Iceland'. *Journal of Economic Behaviour and Organization* **30**: 1–23

Jackson, E.L. 1982. 'The Laki eruption of 1783: impacts on population and settlement in Iceland'. *Geography* **67**: 42–50.

Jennings, A.E., S. Hagen, J. Harðardóttir, R. Stein, A.E.J. Ogilvie and I. Jónsdóttir. 2001. 'Oceanographic change and terrestrial human impacts in a post A.D. 1400 sediment record from the southwest Iceland shelf'. *Climatic Change* **48**: 83–100.

Jónsson, J. 1994. 'Fisheries off Iceland, 1600–1900'. *ICES Marine Science Symposium* **198**: 3–16.

Jónsson, S. 1983. 'The Icelandic Fisheries in the Pre-Mechanization Era, c. 1800–1905: Spatial and Economic Implications of Growth'. *Scandinavian Economic History Review* **31**: 132–50.

Karlsson, G. 2000. *Iceland's 1100 Years*. London: Hurst

Larsen, G. and J. Eiríksson. 2008. 'Late Quaternary terrestrial tephrochronology of Iceland – frequency of explosive eruptions, type and volume of tephra deposits'. *Journal of Quaternary Science* **23**: 109–20.

Magnússon, Á. and P. Vídalín. 1940. *Jarðabók Árna Magnússonar og* Páls Vídalíns Copenhagen: Íslenska fræðafjelagi.

Morrison, S.J.L. 2012. 'Vestfirðir, and the Emergence of Early Fishing Communities in Pre-Modern Iceland'. Unpublished Ph.D. Thesis, University of Stirling.

Ogilvie, A.E.J. 1981. 'Climate and Economy in Eighteenth Century Iceland', in C. Delano Smith, and M. Parry (eds) *Consequences of Climatic Change*. University of Nottingham: Department of Geography. pp. 54–69.

— 1984. 'The Past Climate and Sea Ice Record from Iceland, Part 1: Data to A.D. 1780'. *Climatic Change* **6**: 131–52.

— 1986. 'The Climate of Iceland, 1701–1784'. *Jökull* **36**: 57–73.

— 1991. 'Documentary evidence for changes in the climate of Iceland, A.D. 1500 to 1800', in R.S. Bradley and P.D. Jones (eds) *Climate Since A.D. 1500*. London: Routledge. pp. 92–117.

Ogilvie A.E.J. and I. Jónsdóttir. 2000. 'Sea Ice, Climate, and Icelandic Fisheries in the Eighteenth and Nineteenth Centuries'. *Arctic* **53**: 383–94.

Oslund, K. 2002. 'Imagining Iceland: narratives of nature and history in the North Atlantic'. *British Journal for the History of Science* **35**: 313–334.

— 2004. '"Nature in league with man": conceptualising and transforming the natural world in eighteenth-century Scandinavia'. *Environment and History* **10**: 305–25.

Robinson, R. 2009. 'The Fisheries of Northwest Europe, c.1100–1850', in D.J. Starkey, J.T. Thór and I. Heidbrink (eds) *A History of the North Atlantic Fisheries Volume 1: From Early Times to the Mid-Nineteenth Century*. Deutsches Schiffahrtsmuseum: Bremerhaven. pp. 127–71

Simpson, I.A., O. Vésteinsson, W.P. Adderley and T.H. McGovern. 2003. 'Fuel resource utilisation in landscapes of settlement'. *Journal of Archaeological Sciences* **30**: 1401–20.

Stephensen, M. 1806. *Eftirmæli átjándu aldar eptir Krists hingaðburð, frá ey-konunni Íslandi*. Leirárgörðum: Íslands opinbera vísinda-stiptun.

Thór, J.T. 2009. 'Icelandic Fisheries, c.900–1900', in D.J. Starkey, J.T. Thór and I. Heidbrink (eds) *A History of the North Atlantic Fisheries Volume 1: From Early Times to the Mid-Nineteenth Century*. Deutsches Schiffahrtsmuseum: Bremerhaven. pp. 323–49.

Vasey, D.E. 1991. 'Population, Agriculture and Famine: Iceland, 1784-1785'. *Human Ecology* **19**: 323–50.

— 1996. 'Population Regulation, Ecology and Political Economy in Preindustrial Iceland'. *American Ethnologist* **23**: 366–92.

~ 8 ~

Modernity and Marine Environmental History: Cultural Influences on the Mechanisation of Georges Bank Fisheries, 1905–1912

Matthew McKenzie

May 1912 witnessed what might have been the most unusual federal fisheries debate the US House Committee on Merchant Marine and Fisheries had ever heard. Before the committee sat H.R. 16457, dubbed 'the Gardner Bill' after its House champion Rep. Augustus P. Gardner of Massachusetts, which called for, in essence, a ban on otter trawling (also referred to as 'dragging') in American waters. Initiated by fishermen themselves – from both the hook and line and trawler fleets – this move represented one of the first and last attempts that the fishing community put forward to enact meaningful, positive legislation restricting the mechanisation of fisheries on Georges Bank. Certainly, that made this bill unusual in the history of United States fisheries management up to that time. In addition, however, Gardner's hearing room strategies made this debate stand out even more. Unlike previous hearings that featured only fishermen, scientists and industry representatives, Gardner argued his case through broader, cultural terms that he hoped would sway the committee members' emotions, as well as their intellects, to his favour. Gardner showed committee members film footage depicting how new forms of fishing mangled and mutilated fish landed from Georges Bank. Gardner brought in a witness from the Scottish North Sea fishery to testify how steam trawling with otter trawls had destroyed local communities of fish and fishermen alike within a generation in Europe. The Massachusetts representative also called upon steam trawlers' officers and crew to testify – even against their own immediate economic interests – to the gear's destructiveness and to provide their views on the future wellbeing of the Georges Bank fishery if otter trawling was allowed to proliferate. Gardner even called upon James Brendan Connolly, a prolific fisheries adventure novelist, to encapsulate not just the economic costs of dragging, but also the social and the cultural consequences of allowing otter trawlers to expand on Georges.

Gardner's tactics represented more than just innovative hearing-room argumentation. Examined through the lens of recent discussion regarding the passing of modernity, the Gardner Bill pitted, in stark contrast, two different, but related, ways in which people understood their relationship to the marine environment. First, Gardner sought to make his case not just through the cold and calculated economic and industrial rationalities typical of modern western society's early twentieth-century natural resource management debates. Rather, Gardner equated with those understandings both emotional and subjective perceptions of the ocean. In doing so, the bill's debate highlighted the fact that fisheries regulations affected more than the dollars and cents, stock size and catch efficiencies of the New England fishing industry. The debate also rested upon, and fundamentally influenced, how people culturally perceived the marine realm in and of itself (Ermarth 2011, Gupta 2005, Beck 1992). Ultimately, the 1912 Gardner Bill hearings pitted antiquated, anti-modern, holistic understandings of the ocean against an extension of modernist analytical and calculating studies of the ocean in the struggle to manage one of the world's most abundant fishing grounds. Conducted on the eve of the wholesale mechanisation of the Georges Bank fishery, this debate tells us much about the struggle between modern and anti-modern approaches to fisheries management. In doing so, it also highlights how the marine realm, and maritime people, have challenged and resisted Western modernity's attempts to bring them into its fold.

As marine environmental history has coalesced into a distinct field of inquiry through the past decade, its contributors have produced a wide array of insightful, radically interdisciplinary and powerful scholarship. Exploring human marine resource use and abuse over vast spaces and across long periods of time, one theme in particular stands out: that human beings have done a poor job of preserving and sustaining the marine resources on which they rely. This is hardly a surprising conclusion, given that our finite ability to gauge environmental change under the oceans has emerged only recently (Rozwadowski 2002, Smith 1994, Hubbard 2006). In addition, given that most marine environmental histories focus upon those most immediately tied to marine resource use and management – fishermen, scientists and regulators – the declensionist tendency within marine environmental history is even more understandable.

There is a problem, however, in this pattern. By focusing most intently upon the fisheries scientists, fishermen and regulators, most marine environmental histories perpetuate a vision that fishing is separate and distinct from the wider societies – and their cultural dynamics – that supported the fleets. If the only people – literally the only people – involved in marine resource use were fishermen, scientists and regulators, world fish stocks would be quite healthy. Similarly, if overfishing represented merely a problem of economics and science, then its end could be readily (if not easily) affected. Fishing is not so limited, however. The attitudes, actions and understandings of many people beyond the above-mentioned triumvirate carry significance for the fisheries. Consequently, the focus upon fishermen, scientists and regulators leaves unasked a much larger and more troubling question. What is it about Western societies that has made, and continues to make, marine resource degradation, if not inevitable, then at least so pervasive?

This tendency is especially salient in discussions of the collapse of the Georges Bank fisheries in the late 1980s (Backus and Bourne 1987, Dobbs 2000, Kurlansky 2008, Greenberg 2010). By examining only the fishermen, scientists and regulators involved, the conventional story of Georges Bank's collapse begins with foreign fleets' fishing intensification in the 1960s. The ensuing push for the removal of those fleets often follows in these narratives, a push culminating in the 1976 US–Canadian declaration of the 200-mile Exclusive Economic Zone. Finally, government-induced over-development of both Canadian and American fleets, culminating in the collapse of Georges' groundfish fisheries in the late 1980s, provides the climax and the moral of the story: if only fishermen were less greedy, scientists more assertive and regulators more courageous, we could have avoided this disaster.

What fail to emerge in this conventional narrative are questions concerning the cultural role Americans ascribed New England's Georges Bank fishery itself. Why was it important for Americans – far from New England's shores – to support a foreign policy bent on expelling foreign vessels from Georges Bank? By extension, why was it important for the rest of the nation that New England vessels dominate the Georges Bank fishery? Why did politicians see it as important to maintain and expand New England's small-scale fishery? These questions, hitherto answered only through discussions of industrial development and Cold War politics, share a deeper, longer-running cultural connection – one that peeks out in the Gardner Bill hearings. How fishermen saw their work, how others saw their work, the meanings ascribed to that work and the challenges posed by taking fish from Georges Bank resonated far beyond the vessels, piers, auction houses, research stations and hearing rooms most directly tied to New England's fishing industry.

In other words, and what this chapter seeks to demonstrate, cultural value systems – almost as much as economics and fisheries science, perhaps – shaped and conditioned the ways and intensities in which fishermen took fish from Georges Bank long before foreign factory trawlers arrived, following World War II. In addition to the economic, ecological and scientific questions Georges Bank's fisheries posed in both 1912 and 1976, the fisheries, in part, defined American and New England identity, reflected American attitudes on natural resource use and abuse and, finally, questioned industrial natural resource extraction. In short, before the arrival of foreign factory trawlers in the 1960s, before the push to rid Georges Bank of foreign interlopers and before the ensuing push to modernise and expand New England's 'traditional' industry of fishing, people had to see the fisheries as carrying cultural significance. The fishermen pushing to ban otter trawling in 1912 knew this and, indeed, these understandings shared centre stage with catch statistics and economic analyses in their efforts to prevent Georges from becoming one more industrial 'waste-sea'.

Any invocation of modernity necessitates some definitions and clarifications. I am not interested in taking a position within current debates between continental philosophers and analytics over the next epoch in Western society. Whether we are post-modern, non-modern, 'altermodern' or in the next refinement of the project of modernity is not something I wish to talk about here (Sim 2010). Furthermore, I wish to differentiate – albeit problematically – 'cultural modernity' from the early twentieth century aesthetic movement of 'modernism'. The two are certainly related. Like the authors I will discuss below, however, I am more concerned with value systems, epistemologies and undergirding intellectual assumptions that for the past half millennium had distinguished modern Western cultures from their predecessors. Like Miriam Wright, I see something inherent in the way modern peoples have related to, utilised and manipulated marine resources that has not yet been given the scholarly consideration I believe is warranted (Wright 2000). Finally, I will use the term 'anti-modern' to describe those who oppose the actions and arguments their opponents developed through distinctly modern reasoning. For the anti-moderns I will discuss below, how their opponents saw Georges Bank, used Georges Bank and justified their plans for Georges Bank all represented something fundamentally wrong, ill-informed and only partially considered.

That said, I choose a definition of modernity that has emerged from current discussions on what cultural epoch Western society is entering. If the name of that next period is in dispute, that we are entering something new is not and, in those discussions, theorists have defined Western modernity through a number of features differentiating it from previous social and cultural eras. For Elizabeth Ermarth, Western Europe's modern condition emerged in the late Middle Ages with the intellectual conceptualisation of objectivity, and by extension the creation of a neutral point from which observations could be achieved as objective (Ermarth 2011, pp. 3–29). For many other scholars, modern society is marked by a reliance upon science and technology to unlock nature's potential to free humanity from undeserved poverty and scarcity under the mantle of progress (Beck 1992, p. 19). Ulrich Beck in turn, acknowledges that this system has not guaranteed freedom and liberal societies, but rather emerged to give those claiming scientific or technical expertise 'grotesquely inflated and unconditional power' (Lash and Wynne: 6). Similarly, by the twentieth century, as Stuart Sim argues, modernity had emerged in the West as a social and cultural system wedded to Enlightenment visions of rationality, science and technology.

Through these discussions, a number of common threads emerge which I use as a functional definition of modernity. First, modern society maintained faith in human ability to know nature through detached, systematic and putatively objective scientific observation. Consequently, science emerged as the only and best way to know nature, and did so while eschewing the bias and interested nature of opinions. Second, modernity was marked by the privileged place people accorded instrumental reasoning. Planning and projection defined how people engaged with a wider world, often limiting scope and perception to the terms and issues to be settled by the project and plan themselves. Thus, unintended consequences, wide-ranging effects manifesting themselves far away and undesirable by-products of human activity all could be rationalised away (another important aspect of modernity) as externalities that should not be used to critique the success or failure of planned actions. Finally, the separation of humanity from the superhuman – God, to put it another way – placed modern people in sole physical and moral control of the world they could manipulate for their own exclusive benefit through science. As will be seen below, Gardner's supporters, in marked ways, questioned these forms of thinking and balked at the consequences they saw as inevitable if otter trawling was allowed to expand on Georges Bank.

In 1912, when Augustus Gardner proposed his bill calling for the banning of the importation of fish caught by beam and otter trawls, he likely recognised that modern metrics – in particular the economic and industrial gains otter trawling promised – were stacked against him. In 1905, Boston-based Bay State Fishing Company launched the first steam-powered otter trawler owned in the US, *Spray*. After initially running at a loss, *Spray* soon increased landings, enhanced profits and sparked the firm's growth into one of the most powerful fishing operations in the US. As a result of *Spray's* success, the Bay State Fishing Co. began building two more steam trawlers, while, at the same time, investors in New York organised the *Heroine* Co. in 1911 for the purposes of building a steam trawler to operate out of that port.

While otter trawling was relatively new in North America, the North Sea had supported steam powered otter trawlers since the 1890s. Not only was draggers' efficiency well known in North America, so too was the vessels' ability to drive down European fish stocks. In response, Canadian fishermen successfully pressed for a national ban on otter trawling in 1909 and, along similar lines, the Gloucester Board Trade, a group of Gloucester and Boston hook and line fishermen, began working to enact a similar ban on otter trawling in American waters. In 1911, Canadian fishermen went further, pushing the Nova Scotian House of Assembly to request that Ottawa discuss with Americans ways to prevent otter trawling on the north-west Atlantic continental shelf. In response, Gloucester hook and line fishermen approached their representative, Augustus P. Gardner, to sponsor a similar bill in the US. Thus, by 1912, it was clear that the lines on otter trawling had been drawn (US House of Reps. 1912, pp. 4–20).

Otter trawling's opponents faced significant legal problems. First, most steam trawling took place beyond the three-mile territorial seas over which each state retained regulatory authority and, beyond three miles, no governing body had the authority to regulate fishing. In addition, previous attempts at banning specific fishing gear in US waters had met with only mixed success. Federal-level efforts failed to curtail the use of purse seines in the mackerel fishery in the 1885 Mackerel Bill and states had to fight a rear-guard legal action all the way to the Supreme Court against menhaden fishing companies who questioned states' authority to regulate coastal fishing (Pillsbury 2009, McKenzie 2012). Thus, steam trawling's opponents faced uncertain prospects in any attempt to secure a federal-level ban on any specific gear in any waters. There was some hope, however. While the 1885 Mackerel Bill did not curtail the use of purse seines, proponents of restricted fishing did develop a novel way establishing a *de facto* ban on gear at a federal level. Purse seine opponents had argued that, because the Constitution placed international trade under federal authority, the federal government had the power to ban the importation of fish caught with certain gear on the high seas. This argument gave Gardner and his allies a promising tool. Given the limited time fishermen had to land their catch for the fresh fish market (the destination for almost all fish caught with otter trawls), a ban on the importation of fish caught with otter trawls would not only effectively ban otter trawls, it would do so in both state and international waters. With such a strategy in mind, Gardner shepherded his bill through the US House and it landed before committee on 28 and 29 May 1912.

From the start it was clear that this would prove an unusual fisheries hearing. To open the proceedings, Gardner used the new technology of motion pictures to provide for the committee visual images of how an otter-trawling crew hauled back, dumped, culled and sorted their catch. It also showed the committee how they tossed the rejected fish back overboard. Filmed aboard *Spray* the previous March, the film showed hundreds of fish dumped out on the deck. Most fish were already dead, their bodies distended by burst swim bladders ruptured by the dramatic change in pressure caused by the rapid hauling-in of the net. Committee members then watched as fishermen tromped through the fish in heavy sea boots to save those groundfish in good enough condition for market, while pitching overboard with large forks and shovels those fish too small or too mangled to have any market value at all. They watched, finally, as fishermen pushed scores of fish with no market value – by-catch – back into the sea dead, their sacrifice serving no human need other than to make fishing more efficient and profitable. For a bill seeking to ban otter trawling as wasteful and destructive, Gardner hoped the film would fix in members' memories the reality of mechanised fishing.

The images the committee saw made the impact Gardner had hoped. Committee chair, Joshua Alexander, in discussing the wastefulness of otter trawling, stated 'I saw that [motion] picture showing that when the net came up it looked to me as though 50 or 60 or 70 per cent of the fish were small and all were thrown back into the sea ... What impressed me the most ... [in the moving] picture was the awful destruction of small and young fish' (US House of Reps. 1912, p. 35). In fact, it was trawling's wastefulness that appalled Gardner's allies the most. For these fishermen, such wastage carried more than industrial or economic significance. Indeed, most witnesses had been, or were still, employed aboard draggers and, as such, defied their economic interests in highlighting the waste inherent in that fishing method. But most accounts of dragging's waste carried a moral tone similar to that of William Main, a Scottish fisherman called upon to testify on the effects of trawling on the Scottish North Sea fisheries. When asked what he did aboard an otter trawler with haddock smaller than the eight-inch market minimum, he replied, 'I may say with shame I never did it before, but I was the man put them overboard. I never did it before.' 'You shoveled [the juvenile haddock] overboard?' 'Yes, sir' (US House of Reps. 1912, p. 56).

American captains followed Main's shame and indignation at otter trawling's useless destruction of natural bounty and, in particular, its destruction of immature fish. Captain Francis Carroll testified, for example, 'I have seen as many as 20,000 pounds of fish caught in one haul, and one-third of them would be thrown overboard. All that were thrown overboard were dead.' Captain William G. Thompson, who fished aboard an otter trawler for five trips in 1908 and 1909, recounted, 'on trips where we dragged those nets[,] one-third of our fish would be unmarketable and would be thrown off ... From the way I have seen it in my experience, about 30 per cent of the fish that we catch we destroy ... and they were small edible fish' (US House of Reps. 1912, pp. 67 and 69). Writer and reporter James Brendan Connolly put it most succinctly: 'The thing that struck me most forcibly in the North Sea fishery was the wastefulness of the [otter trawling] methods ... That is above all else that struck me. Those little fish that they throw away, if allowed to remain and get their full growth, would reach several times the bulk in amount of fish that are saved.' (US House of Reps. 1912, p. 77)

Waste carried broader implications, too. The fishermen testifying saw very clearly the ecological threat otter trawling posed. When asked what effect dragging would have upon the Georges Bank fisheries, Thompson replied, 'I think in seven or eight years we would have no fish; they would all be destroyed' (US House of Reps. 1912, p. 71). Carroll was a little more optimistic – he said that 'In 10 years from now I do not believe there will a fish [on Georges Bank]' (US House of Reps. 1912, pp. 64–66). When questioned, he affirmed his belief in the gear's destructive force: 'You estimate that you kill more small fish with this process than you do with the hook?' 'Yes, sir; twice as many. You catch everything on the bottom with that ... you could not get a better thing to do away with the fish on Georges Bank.' Indeed, even committee chair Alexander was led to conclude, 'I suppose if the only problem were to get all the fish out of the sea, the more [otter- and] beam trawlers the better' (US House of Reps. 1912, p. 43).

More was at stake than just stocks of fish. Unlike modernity's tendency to render labour a commodity with a larger social and cultural context, many in the hearing room saw fishermen's work as carrying important moral value to the nation – a value imperilled by otter trawling's changes to the nature of fishing work. According to Fitzgerald, hook and line fishermen typically found in Gloucester, and by extension all other independent fishing captains, represented a breed of American worthy of special consideration: 'I think we ought to favour the Gloucester fishermen. I do not think there is any work that is quite as hazardous, and we ought to help them in every way. Those men go calmly out to sea and many are lost ... thousands of lives have been lost in the fishing business, and yet that business has been closer to American seamanship than anything else.' (US House of Reps. 1912, p. 22) As William Main argued, otter trawling threatened the independence and livelihood of these deserving workers: 'fishermen along the east coast of Scotland as a rule were their own masters and had their own property, and since this trawling came in force and steam fishing came in they have been compelled to leave their own homes to go to seek a livelihood elsewhere' (US House of Reps. 1912, p. 57). Gardner was willing to concede that otter trawling was more efficient; he himself stated that, 'I am prepared to admit that [otter trawling] is a cheaper method of fishing, [against] which these individual captains and small fishermen cannot compete'. But such efficiencies came at a cost too great for the fishery, and another, older, route promised a more respectful handling of the marine resource. 'All we can show by witnesses is that [otter trawls] do throw masses of small fish overboard, which would not be wasted in the kind of fishing which our folks carry on' (US House of Reps. 1912, p. 60). For Gardner's allies, mechanised dragging threatened the livelihood of essentialised, iconic fishermen, whose daily struggle with the sea bred within them unique qualities and entitled them to special consideration.

Gardner faced an epistemological problem with this case, however. From the start, the debate focused on different ways of knowing the marine ecosystem. In contrast to his opponents, who saw statistics, scientific testimony (albeit unstable – more below) and industrial rationales as the foundation for truth, Gardner saw truth as posing some fundamental interpretive problems: 'the fact is, Mr. Garcelon, that some of these reports would be interpreted in different ways by different men, according to whichever side their interests are on' (US House of Reps. 1912, p. 96). Given this, while Gardner was willing to cite and utilise scientific findings and industry statistics when useful, he also based his argument upon emotive and subjective ways of knowing. Clearly, his use of film appealed to the committee's impressionable and subjective faculties. More tellingly, however, his supporting witnesses claimed expertise based not on detached scientific analytical perspectives, but from long-term, personal immersion within the fishery and a marine ecosystem that included the people working it. Fitzgerald, in testifying in favour of the ban, established his credibility by highlighting how he grew up in Boston's fishing neighbourhoods, working with Boston fishermen, and that experience gave him insights into fishermen's hardships (US House of Reps. 1912, p. 21). Similarly, Gardner established his fishermen-allies' credibility by asking each witness to declare how long, and in what role, they had worked in the fishery. Thus, from Gardner's perspective, intimate familiarity with the fisheries served as well as a foundation for assessing natural systems as did detached analytical reasoning.

A US House hearing room was not the right place to forward an emotive argument, however. Simply put, modern legal practices, rhetorical strategies and the very economic nature of the issue at hand combined to made Gardner's position vulnerable. Rather than accept fishermen's subjective allegations, as he saw their testimony, Gardner's opponent, William F. Garcelon proposed the matter be settled through scientific study, focusing only on those issues that Garcelon used to argue against restraint (US House of Reps. 1915). The ensuing study explored how otter trawls mangled captured fish and collected data on the gear's effect on juvenile fish. The study did not address, however, any of the issues that Gardner laid at the foundation of his case: otter trawling's effect on fishermen, their communities and fish stocks as a whole.

In the end, Gardner's Bill did little more than establish two ominous precedents. First, it allowed for the unrestrained use of otter trawls on Georges Bank. Second, it marginalised all forms of expertise not predicated upon scientific, economic or industrial bases. In other words, for the rest of the twentieth century, fisheries questions, on Georges Bank at least, would rest in the domain of modernist, detached, quantitative analysis, with no room for subjective, intuitive and emotive understandings. An important voice for restraint was silenced, and Georges Bank would collapse.

It is not my intention to argue that 'modernity destroyed Georges Bank'. Clearly such a statement diffuses responsibility far too widely for what I seek to establish here. But I do want to argue that fisheries issues were seen as part of a larger cultural milieu – one that included modernist assessments of economics and ecology derived from detached observation, but also non-modern qualitative comprehensions derived from subjective inclusion. There is something inherent in the way modern society understands the non-human world that leads to destruction when actions and attitudes become separated from their eventual consequences. The Gardner Bill debate reveals that both modern and anti-modern epistemologies had a place in discussions over marine resource management and that, in drawing from both forms of understanding, fishermen straddled modern exploitative attitudes and anti-modern mandates for sustainability. With the marginalisation of non-scientific, non-statistical evidence, however, the ensuing primacy of modern calculus sealed Georges Bank's fate. Quite simply, there was no admissible reason to curtail fishing. From that point on, and not from the arrival of the factory fleets following World War II, Georges Bank was destined to be overfished.

BIBLIOGRAPHY

Backus, R.H. and D.W. Bourne. 1987. *Georges Bank*. Boston: MIT Press.

Beck, U. 1992. *Risk Society: Towards a New Modernity*, trans. Mark Ritter. Los Angeles: Sage.

Dobbs, D. 2000. *The Great Gulf: Fishermen, Scientists, and the Struggle to Revive the World's Greatest Fishery*. Washington DC: Island Press.

Ermarth, E.D. 2011. *History in the Discursive Condition*. London: Routledge.

Greenberg, P. 2010. *Four Fish: The Future of the Last Wild Food*. New York: Penguin.

Gupta, D. 2005. *Learning to Forget: The Anti-Memoirs of Modernity*. Oxford: Oxford University Press.

Hubbard, J. 2006. *A Science on the Scales: The Rise of Canadian Atlantic Fisheries Biology, 1898–1939*. Toronto: University of Toronto Press.

Kurlansky, M. 2008 *The Last Fish Tale: The Fate of the Atlantic and Survival in Gloucester, America's Oldest Fishing Port and most Original Town*. New York: Riverhead Books.

Lash, S. and B. Wynne. 1992. 'Introduction' to U. Beck, *Risk Society: Towards a New Modernity*,trans. Mark Ritter. Los Angeles: Sage. pp. 1–9.

McKenzie, M. 2012 'Iconic Fishermen and the Fates of New England Fisheries Regulations, 1883–1918'. *Environmental History* 17(1): 3–28.

Pillsbury, E. 2009. An American Bouillabaisse: The Ecology, Politics and Economics of Fishing around New York City, 1870–Present. Unpublished doctoral dissertation, Columbia University.

Rozwadowski, H. 2002. *The Sea Knows No Boundaries: A Century of Marine Science Under ICES*. Copenhagen: International Council for the Exploration of the Seas and University of Washington Press.

Sim, S. 2010. *The End of Modernity: What the Financial and Environmental Crisis Is Really Telling Us*. Edinburgh: Edinburgh University Press.

Smith, T.D. 1994. *Scaling Fisheries: The Science of Measuring the Effects of Fishing, 1855–1955*. Cambridge: Cambridge University Press.

US House of Representatives. 1912. *Hearings Before the Committee on The Merchant Marine and Fisheries, House of Representatives, on H. R. 16457 Prohibiting the Importing and Landing of Fish Caught by Beam Trawlers*. Washington DC: Government Printing Office.

US House of Representatives. 1915. *Otter-Trawl Fishery: Letter from the Secretary of Commerce Transmitting Communication from the Commissioner of Fisheries Submitting a Report on the Otter-Trawl Fishery* ... Washington, DC: Government Printing Office.

Wright, M.C. 2000. *A Fishery for Modern Times: The State and the Industrialization of the Newfoundland Fishery, 1934–1968*. Toronto: Oxford University Press.

~9~

Producing Fishing Territory and Mapping Ghosts

Robert Snyder

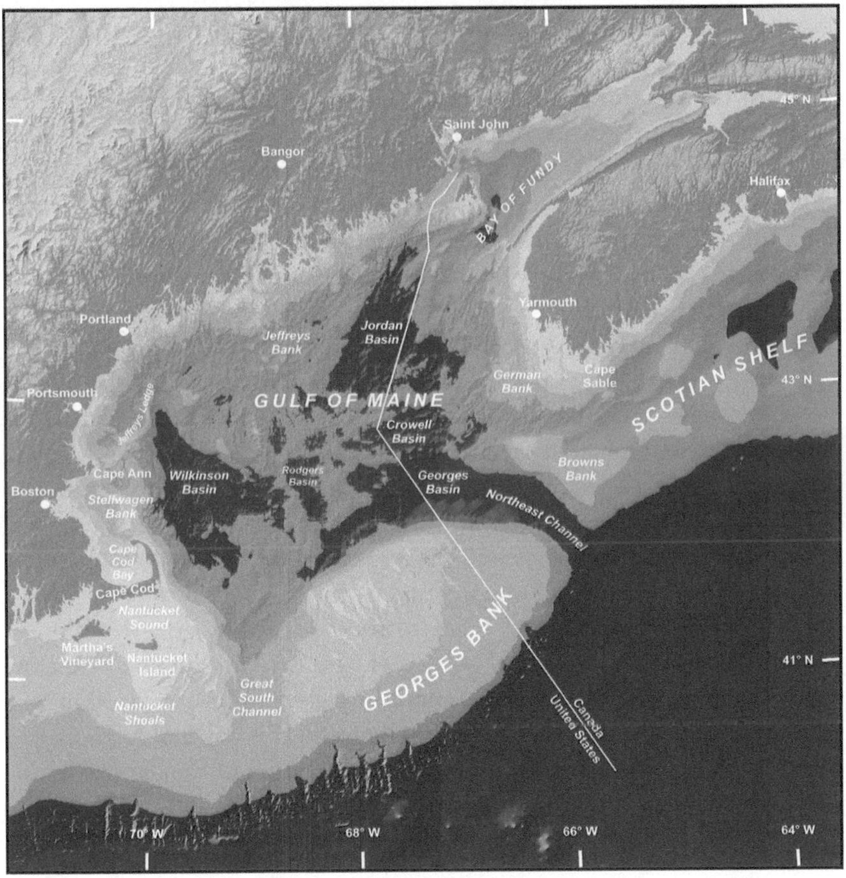

Figure 1. The physical characteristics of the Gulf of Maine www.nefmc.org June 2009. Slash marks demarcate Maine state boundaries.

INTRODUCTION

The abundance of sea fish [is] almost beyond believing.

A Minister in Salem, Massachusetts in 1629,
cited in Kurlansky 1998, p. 70.

In this chapter, I show how bioeconomic thinking, and an assemblage of relationships, events and fish, played a role in producing the New England fisheries management territory. It is a story of making fish technical and of framing fishers as individual exploiters requiring management. The stories of New England fisheries brought to light in this chapter begin to outline claims to territory for some fishers and exclusions from fishing for others. By documenting the work of producing this territory, I lay a foundation for discussing the politics of seeking finer-scale fisheries management territories in New England. I present entangled stories, official narratives and the health of fish stocks, in order to highlight the sedimented history of this area of the ocean. Of particular interest, this chapter illuminates a set of struggles by scientists, government agencies and fishers for authority over the purpose and mechanics of what would become the New England fisheries management region. This discussion is important to outline because it provides context for understanding some of the assumptions that underpin the discursive and material practices that will ultimately be shown to enrol and/or exclude particular management agendas.

The chapter begins by outlining how management decisions and fisheries stock-assessment science converge to produce the region. I then reflect on stories of loss and longing told by fishers about how their fishing communities are disappearing. These stories of loss and longing are followed by an exploration of the federal politics of how the New England fisheries region came into being, what assumptions informed its constitution as a space of authority and what sort of momentum this cultural form carries with it that excludes alternative visions for the future of the region.

COUNTING FISH[1]

> A population [of people] can be counted; but who knows how many fishes are in the sea? And yet it appears to me a project big with possibility, to regard the discoveries of fishery research from a similar standpoint to what has been adopted in the science of vital statistics.
>
> Johan Hjort 1907 (Smith 1994, frontispiece)

Gauging the ocean's bounty is difficult. Nevertheless, Hjort and others are compelled by this challenge of enumerating something that is highly mobile and beyond sight, below the surface of the sea. The thought of this challenge reminds me of the comments a fisheries scientist made in the documentary *Fishing for the Future* while looking out the door of a prop plane, flying low over the treetops of Maine. To paraphrase, imagine that, in order to learn about the health of a particular species of bird, I threw a net out of this plane and dragged it through the forest. We would do this again and again – a relatively small net in a massive forest – and we would extrapolate, from what we drag up, whether or not specific bird populations were healthy.

Historically, discussions of fish stocks focus on the number of fish in the sea, their range, recruitment[2] and resilience.[3] While focusing on fish biology and ecology, these discussions also territorialise fish within a particular ecology, bathymetry[4] and space of authority that enables market-based management of the commons.

In New England, fish are counted a number of ways. The primary data derive from where they are landed (landing port). This is usually the wharf or other infrastructure where the fish are sold to buyers or dealers who then truck them to processing plants or fresh-fish markets. A second means of assessing the health of fish stocks in New England is the federally-sponsored trawl survey. The trawl survey is conducted by a research vessel,

1 St. Martin (2001) shows in great detail how New England fisheries management was shaped by neoliberal approaches to conservation and economy.

2 See note 6.

3 Capacity of a natural system (fisheries community or ecosystem) to recover from a heavy disturbance such as intensive fishing http://www.fao.org/fi/glossary/default.asp (June 2009).

4 Bathymetry refers to the surficial geology of the ocean floor – the 'bottom type'. These analyses are significant because they show how particular areas of the ocean floor are composed of sand, gravel, mud or mixtures of sedimentation that correlate to particular fish habitats at different stages in their lifecycles.

with scientists on board, who count and measure fish. They randomly sample the entire New England region. Another means of gathering data is through random 'at-sea' monitoring whereby fishers must take National Oceanic and Atmospheric Association (NOAA) monitors on fishing trips to observe their harvests. Another data set can be found in each state's 'in-shore' trawl surveys that take place in state waters (between the shore and three miles from shore). The data from each of these surveys are compiled for statistical areas that the National Marine Fisheries Service (NMFS) has created (See Figure 2).

As a result of these survey techniques, detailed by Smith (1994), there is no question that landings data produce an overflow of information relative to the federal and state trawl-survey data and at-sea monitors.

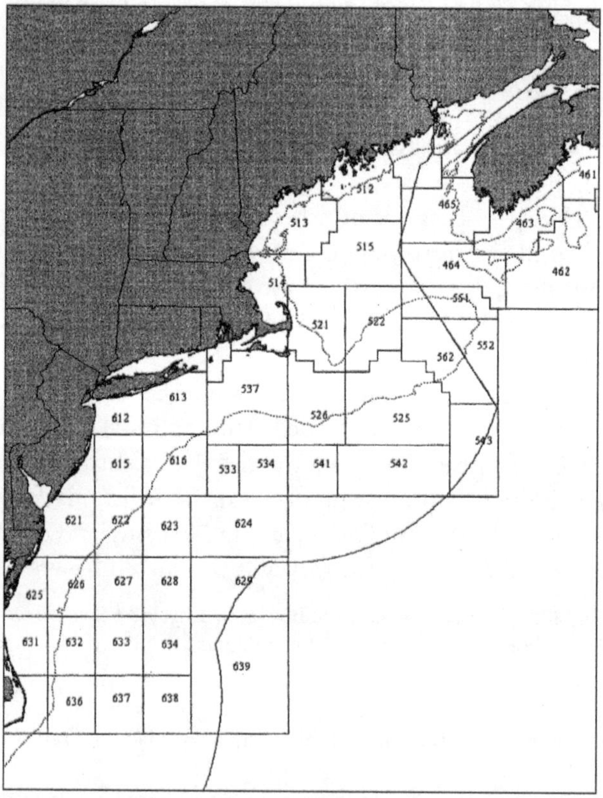

Figure 2. National Marine Fisheries Service, Northeast Fisheries Science Center Statistical Areas. 500s indicate New England, 600s indicate the mid-Atlantic (From St. Martin 2001, p. 126).

The practice of recording fish landings limits our understanding of stock dynamics over time and space because they are based on the movement of fish as a commodity rather than looking to additional sources of information over a specific period of time, such as fishers' historical knowledge or log books where fishers write down the characteristics of their catch.[5]

Historical records of fish landings are generally noted by landing country as shown in Figure 3.

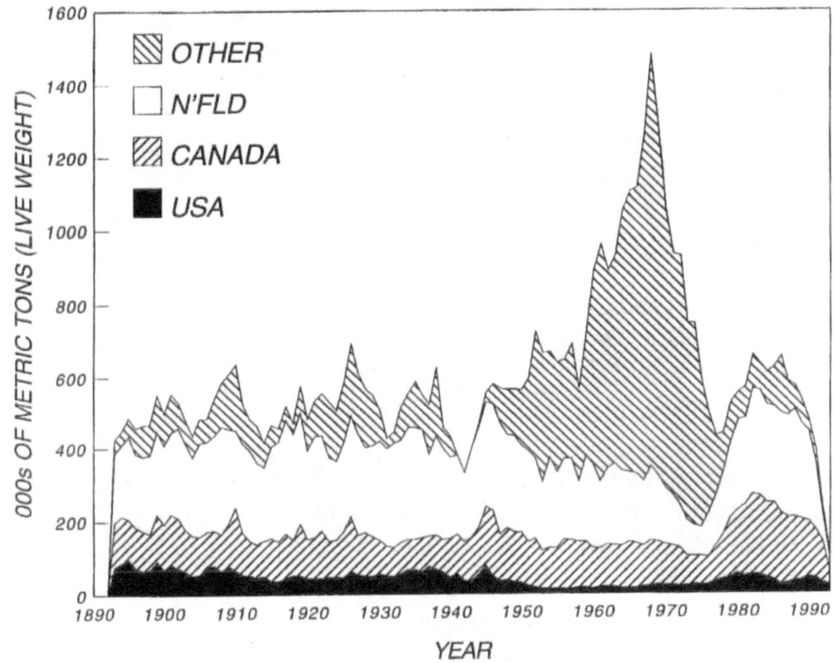

Figure 3. Landings of Atlantic Cod (thousands of tons) by country 1893–1993. Newfoundland retained dominion status until 1949, and its landings are separated from the rest of Canada. 'Other' denotes landings by non-North American Fleets (Murawski et al. 1997).

5 Andrew Rosenberg, professor of natural resources and earth and ocean sciences at the University of New Hampshire, is attempting to remedy this lack of additional finer-scale data by going back through captains' logbooks over the last century (Alexander et al. 2009). Many environmental groups are interested in the results that will be produced because of what they might say about nature prior to the introduction of various fishing technologies and management strategies.

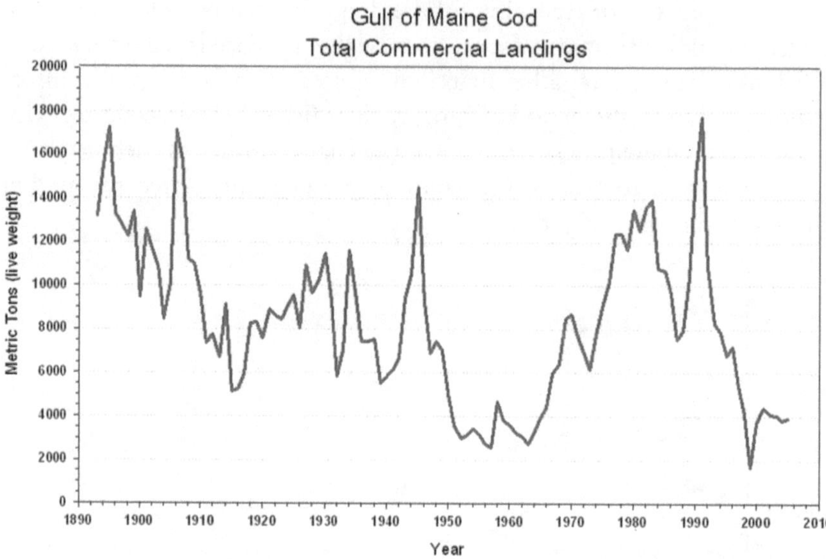

Figure 4. Total Commercial Landings of Gulf of Maine cod (NAFO Div. 5Y) 1893–2005. (Northeast Fisheries Science Center www.nefsc.noaa.gov/sos/spsyn/pg/cod/ accessed January 2010). Gulf of Maine cod is a subset of the entire Atlantic cod range identified in the previous figure.

These two figures show how fishing effort increased dramatically, starting in the 1960s, and continued to climb through the 1970s. It highlights North American fishing effort relative to 'other' fishing effort, indicating how much of the increase in fishing effort came from fishers outside North America. The 'other' identified here consists of fishers from the former Union of Soviet Socialist Republics, Poland, the former German Democratic Republic, the former Federal Republic of Germany, Japan and other countries (Murawski et al. 1997, p. 52).

Stocks that were primarily targeted by Canadian and United States fishers prior to the 1950s were now being exploited by a much larger fleet. During this time cod accounted for roughly fifty per cent of the fish landed out of the North-east. Cod landings peaked in the mid-1960s, then landings of haddock and silver hake peaked over the decade that followed, leading up to the creation of the US Exclusive Economic Zone (EEZ). Estimates of landings from Georges Bank suggest an increase from 240,000 metric tons to 780,000 metric tons in the five years between 1960 and 1965 (ibid.). A second peak in landings followed in the mid 1980s, as the United States fishing fleet expanded to take the place of foreign vessels. In hindsight, it becomes clear that groundfish stocks were being caught at rates that were increasing radically during the 1960s and 1970s. In fact, it is hypothesised that overfishing specifically where recruitment[6] was taking place, decimated the stocks: 'had recruitment overfishing been prevented, catastrophic declines in these resources could have been prevented' (Sinclair and Murawski 1997, p. 90). However, observers and industry representatives during the mid-1960s would have been looking at the overall fishing effort for key commercial species such as cod and seen a curve that expanded endlessly; and they might have assumed that groundfish stocks were limitless.

By highlighting these figures, I seek to show how stock-assessment science based primarily on landings data enables two agendas. First, we see the apparent limitlessness of fish and, second, we see how New England fisheries and, more broadly, North American fisheries, are being exploited by foreign fleets. Data like these would inevitably play into the geopolitical goals of politicians thinking hard about how to exclude foreign fleets from taking the resource, so that they could capitalise a United States fishing fleet to capture the economic value of the apparently limitless amount of fish in the ocean, a story to which I now turn.[7]

ENCLOSURE

The massive increase in 'other' fishing effort after World War II noted in Figure 3 was a cause of major concern among fishers and within the Executive Office of the President of the United States. Spencer Apollonio and

6 Recruitment: the amount of fish added to the exploitable stock each year due to growth and/or migration into the fishing area. For example, the number of fish that grow to become vulnerable to the fishing gear in one year would be the recruitment to the fishable population that year. This term is also used in referring to the number of fish from a year class reaching a certain age. For example, all fish reaching their second year would be age 2 recruits (www.nefsc.noaa.gov/techniques/tech_terms.html#tt12 accessed May 2010)

7 The data tell us what was landed and sold, not what was caught. We only know what made it to market in places where fish were landed for commercial purposes. We do not know what was thrown overboard, nor do we know what was kept for communities and neighbours. This data stream also says nothing about a history of technological innovations and increases in capitalisation of the fisheries that would have increased the capacity of fishers to land more fish.

Jacob Dykstra,[8] now in their eighties, both served during the years leading up to the formation of the New England Fishery Management Council and they were at the council table during the first ten years of its existence. They have taken the time to self-publish their version of these negotiations in *'An Enormous, Immensely Complicated Intervention': Groundfish, The New England Fishery Management Council, and the World Fisheries Crisis* (Apollonio and Dykstra 2008). I draw heavily on their reflections to outline the politics of creating the New England fisheries region.

In their view, World War II had significantly diminished the European fishing fleet, adding to hunger in the region. One strategy that Europeans used to overcome this problem was to commission the construction of new fishing fleets that took full advantage of the sonar technology developed during the war. For the first time in history, ships were built that could come from as far as Japan to fish the stocks off of Canadian and United States shores. These were factory trawlers[9] that could catch, process and freeze fish all on the same vessel. The arrival of these vessels increased fishing effort directed at these stocks of fish by 300 per cent in a mater of a decade (Alverson 2002, p. 5, in Apollonio and Dykstra 2008, p. 89).

In the summer of 1963, over 300 Russian factory trawlers focused on fishing Georges Bank, animating political debates about the threat of communism that were also prevalent at the time. United States Congressman Thomas Eastland commissioned the Eastland Fisheries Survey in 1973 while also maintaining a high profile in efforts to quell communism in the United States throughout the 1950s, 1960s and 1970s. He would have viewed the arrival of hundreds of Russian fishing vessels off United States shores in the 1960s and 1970s as cause for great concern.

Furthermore, as foreign fishing fleets grew in the 1960s and 1970s, policymakers recognised that the United States fishing fleet had become one-sixth the size of Japan's, Germany's and other nations' fishing fleets. They

8 Dykstra has 60 years of experience in fisheries management; he founded a fishing cooperative, and was a member of the United States delegation to the United Nations Conference on the Law of the Sea, playing a key role in the establishment of the 200-mile US Exclusive Economic Zone. Apollonio was the Commissioner of the Maine Department of Marine Resources for over ten years and served as the first executive director of the New England Fishery Management Council.

9 Factory trawlers are large stern trawlers equipped with a plant for gutting, filleting, freezing and storing fish, and for processing fish oil and fish meal. Such vessels usually have extensive superstructures http://www.fao.org/fi/glossary/default.asp (Accessed July 2009).

saw this as a lost economic opportunity both in United States waters and abroad. This recognition would eventually shape the creation of the Fishing Vessel Obligation Guarantee Program in 1976. Under this programme, the United States government, through NOAA, would guarantee 100 per cent of local bank loans made for building boats. It was this fund that capitalised many of the boats built through the early 1980s. Some still fish today along the coast of Maine. This fund was one of a number of federally-sanctioned strategies for occupying the New England fisheries region with a home-grown fleet of fishing vessels.

While consciousness of the impact of foreign fishing fleets was growing in Washington, fishers, who were seeing the impact first-hand, were organising with their congressional leaders to extend jurisdiction over the ocean out to 200 miles, a distance that was agreeable to a number of other countries that believed their fish stocks were under threat from roving fleets. The United Nations (UN) responded to these concerns by holding a series of law of the sea conferences in the late 1950s and early 1960s, meant to take up not only fishing issues but also issues of mineral exploitation and oil drilling in offshore waters.

This process revealed powerful interests in the United States and other countries that were strongly opposed to the institutionalisation of a 200-mile line. According to Apollonio and Dykstra, the military and the mineral agencies were particularly strong in their opposition because of the limitations that such laws would have on their ability to access rich resources and strategic military interests off the shores of other countries.

> The United States Navy and the State Department wished to preserve maximum freedom of navigation. The US High Seas Tuna and Shrimp Industries which fished off of foreign shores were opposed. The US oil and mining industries and their government wished to retain access to mineral resources on and beneath the continental shelves and in deep waters around the world (Apollonio and Dykstra 2008, pp. 93–4).

While it is hard to imagine that this partnering of interests could be defeated in congress, by the early 1970s, Alaska and New England fisheries advocates persevered and were able to find legislative sponsors.

BIOECONOMIC MANAGEMENT

> The goal for domestic fisheries management must be the development of a technically advanced, and economically efficient fishing fleet with the minimum number of units required to take the catch over a prolonged period of time.
>
> Commission of Marine Sciences, Engineering and Resources (1969, p. 93) in Apollonio and Dykstra 2008, p. 91.

As legislation began to take shape around the need to exclude foreign fleets, it became clear that the only way to get presidential support would be to include management as part of the legislation, the purpose of which is quoted above.[10] The addition of management to the legislation under development was cause for concern among fishers. According to participants in the process, this concern came not from a lack of interest in management, but from a disbelief that the United States fishing fleet could possibly be large enough to overfish (Dewar 1983).

Advocates from Alaska stepped forward and proposed a system of regional councils, based on their regional salmon-council model. This proposal was reinforced by research presented in the Eastland Fishery Survey of 1973 that showed the regional nature of fish stocks. This proposal, once adopted, set up a new form of government for United States natural resources, one that handed significant powers to industry councils who would then make recommendations to the National Marine Fisheries Service. The congressional legislation developed did not specifically address how this sharing of power would be handled and so it was left up to the regional councils and NMFS to sort it out. How would these councils manage the newly established regions? What principle would guide them? Bioeconomics.

Bioeconomics is the study of the dynamics of living resources using economic models (Gordon 1954). The United States government, informed by bioeconomic thinking, borrowed key concepts from this field that remain central to fisheries management today. The concept of *maximum economic yield* (MEY) emerged as perhaps the most important to indicating the intent to rationalise fisheries management. MEY is defined as the amount of fish harvested that: (a) will provide the greatest overall benefit to the national

10 According to Apollonio and Dykstra, there was a time when fishing interests almost had an opportunity to have total control over the resource without governmental oversight.

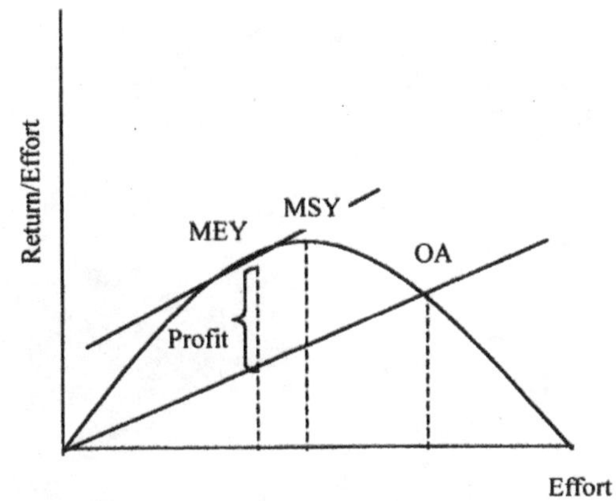

Figure 5. Bioeconomic relationship between maximum economic yield (MEY), maximum sustainable yield (MSY), open-access yield (OA). Source: National Marine Fisheries Service 1996.

economy, particularly with respect to food production and recreational opportunities, and taking into account the protection of marine ecosystems; (b) is prescribed as such on the basis of the maximum sustainable yield from the fishery, as qualified by any relevant economic, social or ecological factors; and (c) in the case of an over-fished fishery, provides for rebuilding to a level that produces the *maximum sustainable yield* in the fishery (see Figure 5). But determining how to manage to MEY would be at the discretion of the councils

In other words, since government agencies understood that overfishing might be taking place, they created MEY and MSY as a possible mechanism for guiding decision-making about how best to regulate effort in the fishery to achieve MEY.[11] They established these concepts as opposed to open-access fishing which implies overfishing and an overcapitalised fleet that leads to declines in catch per unit of effort.

Despite the lack of details about how management of the region would work, on 13 April 1976, the United States Congress passed the Magnuson-Stevens Fishery and Conservation Act of 1976 (known as the

11 If this all seems a bit contradictory, it very well may have been. If congress understood the fish stocks to be overfished, and believed that developing management measures would help control the situation, then why did they create the funds to further capitalise the fishery? But then again, politics in general are often not coherent.

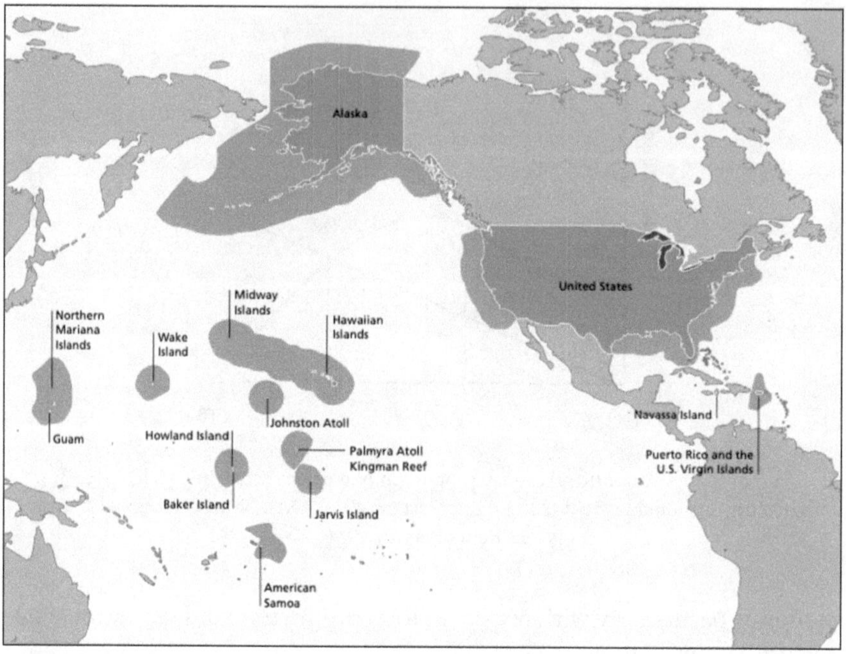

Figure 6. This map depicts the US Exclusive Economic Zone. Downloaded from www.aquaculture.noaa.gov/pdf/20_eezmap.pdf (January 2009, cropped from original).

Magnuson act). With this vote, the United States extended its territory out to 200 miles and created eight regional councils for the purpose of governance.

The intent of the legislation was: (1) removal of foreign fishing vessels, (2) development of the United States fishing industry and (3) conservation and management of domestic fisheries.[12]

In 1983, Margaret Dewar, who had been very involved in drafting the legislation, reflected,

> the legacy of the many years of work to limit foreign fishing and to extend fisheries jurisdiction to 200 miles included much more than the law itself. The experience had molded views among industry spokesmen, fishermen,

12 To say that the original legislation wasn't perfect would be an understatement. Between 1976 and 2009, there were 42 framework changes and 17 amendments initiated through the council process. Opponents of fisheries management often point to the need to manage at the New England fisheries regional scale as the fundamental fault of the council process.

> and officials at NMFS and NOAA that promised to affect the industry and fishery policy in the new era (Dewar 1983).

Thus, the New England fisheries region was born.

Figure 6 shows the extent of this decision across the ocean – over 21 million kilometres of ocean claimed by the United States. The map shows an enclosure of ocean space, an exclusion of access to non-United States fishers, and it highlights the extent of the reach of government over this new territory.

In the next section, I introduce the effects that the enforcement of this territory is having in Maine and introduce ways that fishers have begun to work for finer-scale management territories.

LOSS AND LONGING

Fishermen remember the ocean as a much different place from where they fish today. They recall, albeit with remorse, a place of abundance. Frustrated, they take aim at the New England Fishery Management Council, for, as I will show, within a decade of the New England Fisheries management authority's coming into being in the late 1970s, the fish began to disappear.

I wanted to know what came before: before limited-access permits, bigger boats, the Magnuson act; how did they fish back in the day? I wanted to understand the links between then and now, between the ideas that were being proposed by community-based fisheries management groups in the region and the time before federal oversight. How were these ideas constrained by or enabled by the scales of fisheries management that currently exist?

Sitting down with 'Terrible Roger' Libby (he has this nickname because every time anyone asks how he is doing, he replies, 'terrible') and 'Big Ed' Thorbjornson, I feel as if I am in the presence of giants. The sun is going down over Port Clyde and their sons are out fishing. They sit with cocktails in hand waiting for the phone to ring so that they can go down to the sea, as they have done for decades, to get fresh fish from the boats. An introduction by Roger's eldest son, Glen, and an offering of spirits gets me a seat at the table for a lively evening.

These guys have seen a great deal, having lived by the tides long into their sixties and seventies. Big Ed's family hails from Sweden and he has fished his whole life, getting his start on one of Dick Jellison's redfish boats out of Rockland. Ed lost one of his sons, Gary Thorbjornson, at sea in 1995.

Roger started in trucking – 'worked for Hoffa', he says with a dry humour that keeps everyone but his family and close friends guessing.

Almost anyone who has had the opportunity to sit with lifelong fishers will have heard stories of how it used to be. For example, bringing in 300,000 pounds of redfish a day, at three cents a pound. The economics of the fishery – high volume, low quality, low price –combined with post–World War II technologies, put incentives in place to catch the last fish.

One statistic that clearly shakes Roger and Ed to the bone is that Port Clyde is now the second-largest groundfish-landings port in Maine next to Portland. 'How could that be?' Roger asks, recalling the days when there were active groundfish fleets dotting the coast. Today only one groundfish fisher remains active along the approximately 2,700-mile coastline between Port Clyde and the Canadian border.

For Ed and Roger the story was relatively straightforward. They recall a 'day-boat fishery' right up through the 1970s. 'You would go out, load up [the boat] and return to port over the course of a long day.' They relay how 'fishermen use to play a larger role in their communities and their families back then'. Back when 'there were fish up in close', they fished in state waters, defined as three miles from shore. This meant that they only needed a state permit to fish and, if they went beyond state waters, they were in international waters and could do what they liked. It meant that other countries could fish there too: disgust at the thought of this crosses their faces in the telling.

The economy of Maine's 140 or so coastal towns relied on fish harvested and sold in this day-boat fishery (Wilson 1999). Fishers spent some time fishing for groundfish, some time lobstering, some time clamming – 'a little bit of this and a little bit of that' to support their families. 'We had our territory', Roger tells me. He is referring to informal but fiercely defended boundaries in the water that demarcated where each harbour's fishermen could fish. All of this went away when the fish could no longer be found near shore (Wilson 1999).

For reasons unknown, reasons that fishers and I have debated for hours, the fish stopped coming inshore. 'In Canada they are gone; around here you can still catch a few, but to the southerly they can still fill their boats in a day', referring to New Hampshire and Gloucester fishers who still have a day-boat fishery. Ed is referring to the contradictory idea that the south-westerly ports still have a day-boat fishery despite having large numbers of boats fishing in the area. 'How could it be like that, Rob?' I am sitting with cocktail in hand, at terrible Roger's 76th birthday party, and all I can say at first is, 'Beats me?' I raise the issue of nonpoint source pollution but they move to quickly dismiss this idea – 'pollution; now, Rob, don't you think Boston, Portsmouth and Portland have a hell of a lot more pollution than here? Now that I'm thinking about it, says Roger, maybe that is what we need: more of it!' Not what I expected, but logical in a twisted sort of way. I don't have the courage to voice the thought that rushed to me a few moments earlier… *maybe you just caught them all.*

For these men, the loss of the day-boat fishery is one of *the* great unsolved mysteries; that the fish 'disappeared' even through they had a small fishing fleet, while ports with much larger fleets still have access to a productive fishery.

Contrast this story of the past with the dominant narrative of today's fisheries. Fishers go out on three to five-day trips in their boats – boats that are about the same size as they were in the 1960s and 1970s when they started out – but they are always uncertain that they will be able to land enough fish to pay the bills and feed their families. Fishers are going much farther from shore in order to find fish. The farther offshore they go, the more dangerous it gets in a small boat or the less they fish, because they have to be very careful about ending up out there 'when it's blowin''

MAPPING GHOSTS

Roger's and Ed's experiences are reinforced by historical data captured by Ted Ames in 1996.[13] The basic logic that Ames followed in his historical

13 In 1996 Ted Ames was a fisher out of Stonington, Maine and a former student and researcher at the University of Maine. Through a mix of interviews and mapping technology, Ames was able to learn the historic locations of cod and haddock spawning grounds in the Gulf of Maine (Ames 1997, 2004, 2006). Ames' findings were unprecedented at the time. The fishers he interviewed identified 684,000 acres of haddock and cod spawning grounds; much of this territory was far closer to shore than scientists had previously considered possible. Ames identified what many now believe to be separate spawning stocks of cod and haddock with high natal-site tenacity; meaning that, in much the same way that salmon return to their native rivers, sub-stocks of haddock and cod return to native near-shore spawning grounds. This evidence of finely-scaled and geographically-distinct stocks became one of the key pieces of information used in advocacy efforts to create finer-scaled management in New England.

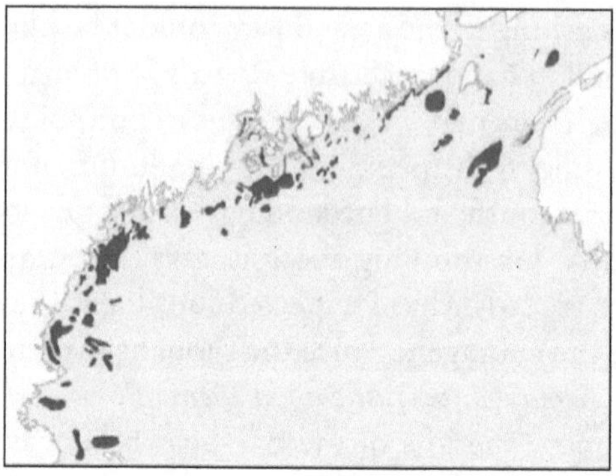

Figure 7. Map produced by Ted Ames showing cod and haddock spawning grounds along the Gulf of Maine coast, identified in the study (Ames, unknown date).

ecology research was that the fishers fished where the fish lived. Map where they used to fish and you will learn the characteristics of stocks in history.

Through time spent sitting with and interviewing groundfish fishers, Ames was able to identify cod and haddock spawning grounds that do not show up on the maps of present- day cod-tagging studies (see Figure 7). This map, like stories of loss and longing, makes a significant contribution to understanding how New England fisheries began to disappear.

It is important to locate Ames' work in a different vein to that of the scientists conducting stock assessments. He is trying to allow people to think about what might be possible in terms of rebuilding stocks, based on fishers' experiences from the past. His work could be interpreted, in part, as a political project, an attempt to use these representations of place to evoke conversation and new thinking.

His work generates a narrative open to discussion among people who are no longer providing the data at ports where fish are no longer landed. Work like Ames's is an attempt to stay on the map of fisheries policy-making while also reminding those who used to fish for groundfish, and the rest of the public, that things used to look far different from how they do today. His is not an appeal to return to pre-contact nature; instead, it is an appeal to think about ecosystem characteristics, to look at the geography and distribution of fish spawning grounds and to think anew about how this ecology might be better served by finer-scale management.

CONCLUSION

If a community has no landings data and no fishers, then the fishers who are still in the fishery trying to make a living (and who dominate the regional management council) view these communities as no longer relevant to management decisions.

The Northeast Regional Cod Tagging Program is interesting for how it reverberates with efforts to quantify movement, how it speaks to loss and longing and how it highlights the challenges of quantifying the behaviours of fish in areas that are highly depleted. In the end, the cod-tagging programme mapped roughly 7,000 reports by fishers who caught and reported a tagged cod. The evidence provided by these tags, based on the geographic location of where the fish was tagged compared to where it was recaptured, demonstrated what fishers told me time and again: cod don't come into the near-shore waters in Maine anymore.

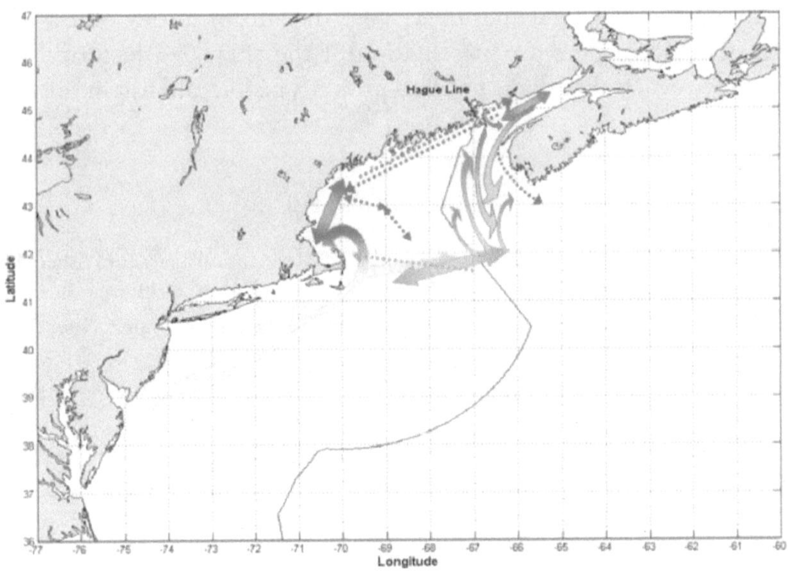

Figure 8. Results of Cod Tagging study identify stock movements throughout New England. The near-shore dotted line approximates the Maine coast and shows the lower signal strength from the data stream that was captured throughout this area (Tallack 2008, p. 15)

Today, the fish that traverse the Gulf of Maine generally move from their nursery grounds in the Bay of Fundy, over the Scotian Shelf, and then they head offshore to the Georges Bank (grey arrows). Another sub-stock of cod originates south of Cape Cod (black arrows) and migrates around the cape to be caught in near-shore waters between Gloucester, Massachusetts and Portsmouth, New Hampshire The number of fish tagged and recaptured in the near-shore waters of Maine was so small that migration patterns couldn't be clearly identified (dotted lines along the Maine coast).

The evidence of finding few cod in the near-shore Maine waters is important for two reasons. First, when this experience is repeated hundreds of times by researchers, you end up with a hole in the map that shows that there are few fish. This confirms what fishers know. Second, if there are no fish, then there are no landings data. An absence of landings data equates with an absence of groundfish fishers. Without fishers, it is difficult to have a voice in fisheries management.

The stories of New England's fisheries brought to light in this chapter begin to tell the saga of claims to territory, exclusions, unintended erasures and attempts to make the invisible visible. This mix of stories, official narratives and fish enables a representation of the entangled historical and discursive artefacts that shape political projects, people and fish in the New England region.

BIBLIOGRAPHY

Alexander, K.E. et al. 2009. 'Gulf of Maine Cod in 1861: historical analysis of fishery logbooks, with ecosystem implications'. *Fish and Fisheries* **10**(4): 428–49.

Ames, E.P. 1997. *Cod and Haddock Spawning Grounds in the Gulf of Maine: From Grand Manan Channel to Ipswich Bay*. Rockland, ME: Island Institute.

— 2004. 'Atlantic Cod Stock Structure in the Gulf of Maine'. *American Fisheries Society* **29**(1): 10–27.

— 2006. 'Reconstructing the Gulf of Maine Cod Spawning Grounds on the Basis of Local Ecological Knowledge. Journal of Maine Biological Assessment' **80**(3675): 1–9.

— Unknown. 'Putting Fishermen's Knowledge First: The Promise and Pitfalls', in www.penobscoteast.org/documents/PuttingFishermensKnowledge.pdf [accessed 1 January 2009].

Apollonio, S. and J. Dykstra. 2008. *An Enormous, Immensely Complicated Intervention: Groundfish, The New England Fisheries Management Council, and the World Fisheries Crisis*. Montgomery, AL: E-Book Time.

Dewar, M. 1983. *Industry in Trouble: The Federal Government and the New England Fisheries*. Philadelphia: Temple University Press.

Gordon, H.S. 1954. 'The Economic Theory of a Common-Property Resource: The Fishery'. *Journal of Political Economy* **62**(April): 124–42.

Kurlansky, M. 1998. *Cod: A Biography of the Fish that Changed the World.* New York: Penguin.

Murawski, S.A. et al. 1997. 'Groundfish Stocks and the Fishing Industry', in J. Boreman, J.A. Wilson and R.L. Kendall (eds) *Northwest Atlantic Groundfish: Perspectives on a Fishery Collapse.* Bethesda, MD: American Fisheries Society. pp. 27–70.

National Marine Fisheries Service. 1996. 'Our Living Oceans: The Economic Status of U.S. Fisheries', in *NOAA Technical Memorandum*, Vol. NMFS F/SPO-22. Washington, D.C.: United States Department of Commerce.

Sinclair, A.F. and S.A Murawski. 1997. 'Why Have Groundfish Stocks Declined?' in J. Boreman, J.A. Wilson and R.L. Kendall (eds) *Northwest Atlantic Groundfish: Perspectives on a Fishery Collapse.* Bethesda, MD: American Fisheries Society. pp. 71–94

Smith, T.D: 1994. *Scaling Fisheries: The Science of Measuring the Effects of Fishing, 1855–1955.* Cambridge: Cambridge University Press

St. Martin, K. 2001. 'Making Space for Community Resource Management in Fisheries'. *Annals of the Association of American Geographers* **91**(1): 122–42.

Tallack, S. 2008. *The Northeast Regional Cod Tagging Program: Addressing Critical Management Questions about Cod Movement and Growth by Ensuring Maximum Tag Returns, Advancing Data Analysis and Incorporating Historical Tagging Data into a Regional, Multi-Program Database.* Portland, ME: Gulf of Maine Research Institute.

Wilson, D.C. 1999. *Fisheries Science Collaboration: The Critical Role of Community.* Hirtshals, Denmark: Institute for Fisheries Management and Coastal Community Development.

III. SCIENCE AND LOCAL KNOWLEDGE

~ 10 ~

From Cook to Cousteau: The Many Lives of Coral Reefs

Alistair Sponsel

INTRODUCTION

Anyone interested in marine environmental history today is likely to feel some concern about the present and future of coral reefs. Living reefs face a range of threats, from direct physical damage and the effects of local pollution to problems presented by deforestation on dry land, which leads to erosion and ultimately to damaging sediment being deposited on reefs. Reef-building corals are threatened too by rising ocean temperatures, which can cause coral bleaching, a phenomenon in which coral polyps lose the alga-like photosynthetic organisms that dwell within them. If that weren't enough, increasing acidification of ocean water weakens corals' mineral skeletons.[1] It is no surprise, therefore, that a great deal of contemporary reef science is aimed at conserving reefs by mitigating the threats posed to them by human activities.[2]

Not long ago, however, scientific curiosity and popular interest in coral reefs were driven instead by the threats that reefs posed to human activity. Far from feeling that reefs needed to be protected, navigators, naturalists and naval administrators in the eighteenth and nineteenth centuries believed that generating scientific knowledge of reef growth might provide some measure of protection to humans undertaking the risks of navigating the tropical oceans or investing in maritime commerce.

This essay aims primarily to recapture that former view, and to show how it changed only gradually over two centuries. In sketching this history of Western attitudes toward coral reefs, I trace a broad shift from

1 On the current threats to coral reefs, see Bryant 1998 and Burke and World Resources Institute 2012. A useful text on corals and their evolution and distribution is Veron 1995.

2 For an overview of the science of reef-management, see Birkeland 1997a.

the view that they were robust and threatening natural phenomena to the view, characteristic of the mid-twentieth century, that they were resilient and benign. In closing I offer a preliminary account of why, in the last few decades, reefs have come to be seen as fragile and threatened.

REEFS AS THE PRODUCT OF CORAL GROWTH

The range of views I examine in this essay all depended on the idea that coral reefs were organic formations. Whereas the notion that corals were animals rather than plants (or minerals) dates to the early eighteenth century, the distinct idea that the whole submarine structure of reefs owed their origin to coral growth was first proposed by Johann Reinhold Forster in the 1770s.[3] A polymath from Polish Prussia who had emigrated to England in 1766, Forster served as naturalist on Captain Cook's second voyage to the Pacific (1772–1775). This was the voyage in which Cook made a series of traverses in an effort to find (or disprove the existence of) a conjectured great continent in the Southern hemisphere. Cook's route provided Forster with several opportunities to observe the central Pacific's distinctive low, ring-shaped islands and reefs that are now known as atolls. These 'low islands' were familiar to Europeans from earlier voyages of exploration, including the 1766–1769 circumnavigation by the French naval officer Louis-Antoine de Bougainville, who had identified a central paradox to their existence. On the one hand, the islands were so slight and flat they seemed on the verge of being swallowed by the ocean, mere 'strips of land that a hurricane could bury at any moment beneath the water'. Of these 'almost drowned' islands, Bougainville wondered, 'Is this extraordinary land rising, or is it in ruins?' On the other hand, the unassuming appearance that made the islands seem potentially ephemeral meant that they were a terrible hazard to navigation, particularly if a ship were making sail in the dark or in bad weather. Discovering a cluster of these islands east of Tahiti, Bougainville named the area *l'Archipel Dangereux*.[4]

Before Forster, it was understood that such reefs offered fertile ground for the growth of corals, whose stony skeletons formed a substance called

3 The best information on Forster is to be gleaned from Hoare 1976 and the introductory essays in Johann Reinhold Forster 1996 [1778].

4 L.-A. de. Bougainville 1772; L.-A. de Bougainville, Bideaux and Faessel 2001. On the Bougainville expedition in the broader context of French exploration in the Pacific, see Moureau 2004.

'coral rock'. This coral rock could be found on the reefs of the low islands and on the extensive shoals that Captain Cook had discovered on the east coast of New Holland [Australia] during his first voyage. Amidst this 'labyrinth' that became known as the Great Barrier, Cook's ship *Endeavour* had run aground despite his men taking every precaution to prepare for suddenly shoaling water. His journal for 11 June 1770 records that a few minutes before 11:00 p.m., the depth was '17 [fathoms (102 feet)] and before the Man at the [sounding] lead could heave another cast [to measure the water's depth] the Ship Struck and stuck fast ... upon the SE edge of a reef of Coral rocks' (Cook 1955, Vol. 1, p. 344). Though he was able to save the ship, it took the *Endeavour*'s company two months to sail clear of the reefs for good (Beaglehole 1974, pp. 236–246).

Sailing on Cook's next voyage, Forster's innovation was to suggest that, rather than merely inhabiting reefs and other shallow areas, corals had created the entire structure of the low islands. Upon viewing the low islands of Takaroa and Takapoto in August 1773, Forster entered into his journal a new account of how animalcules would produce a circular reef.

> Both [of these low islands] have the longer sides [oriented south-north] & the shorter ones [oriented east-west,] ie the long sides are exposed to the East-wind, which probably was the first origin of these curious Isles: for there being some little Elevation or inequality in the bottom of the Sea the Coral-Animals began to raise their large rocky branching out habitations; & as soon as the Animals came nearer to the surface of the Sea, they extended their branches more North, in order to shelter themselves against the Surf of the Sea, & at last this work formed a Lagoon in the middle.[5]

Given the immense depth of the ocean around these islands (Cook did not carry a rope long enough to sound the bottom), Forster was claiming coral growth on a massive scale. He advanced this theory in his 1778 *Observations Made During a Voyage Round the World,* in which he concluded that 'the method employed by the animalcules in building only narrow ledges of coral rocks, to secure in their middle a calm and sheltered place ... seems to me to be the most probable cause of THE ORIGIN of all THE TROPICAL LOW ISLES, over the whole South-Sea' (J.R. Forster 1778, p. 151).

Forster's claim about the role of corals' 'instinct' in building reefs received support from the British navigator Matthew Flinders, whose 1801–1803 survey of the Australian coast led him to declare that the

5 Entry for 18 April 1774. Johann Reinhold Forster and Hoare 1982, vol. 3

geographical extent of 'the reefs, which form so extraordinary a barrier to [northern] New South Wales ... is not to be equalled in any other known part of the world' (Flinders 1966 [1778], pp. 101–02). Along with drawing attention to the reefs' sheer magnitude, Flinders offered a precise account of how successive polyp generations might contribute to these structures. 'It seems to me', he explained, 'that when the animalcules which form the corals at the bottom of the ocean, cease to live, their structures adhere to each other, by virtue either of the glutinous remains within, or some property of salt water; and the interstices being gradually filled up with sand and broken pieces of coral washed by the sea, which also adhere, a mass of rock is at length formed. Future races of these animalcules erect their habitations upon the rising bank, and die in their turn, to increase, but principally to elevate, this monument of their wonderful labours.' Although the reefs Flinders encountered were not the ring-like low islands that Forster had seen, he too saw evidence that coral polyps shared an 'instinctive foresight' to coordinate their behaviour against the power of the wind. He argued that '[t]he care taken to work perpendicularly in the early stages, would mark a surprising instinct in these diminutive creatures. Their wall of coral, for the most part in situations where the winds are constant, being arrived at the surface, affords a shelter, to leeward of which their infant colonies may be safely sent forth.' (Flinders 1966, p. 115)

THE BENEFICENT POWER OF CORALS

The implication of observations by Forster, Flinders and others was that corals potentially could produce major changes to the face of the globe. In the early nineteenth century, this power of mere animalcules to create great physical structures was interpreted in a variety of ways. For those who inclined toward natural theology, the belief that one could come to know God through the study of his creation, corals were agents of providence. These unassuming creatures were a divine means of refreshing the earth with new landforms to compensate for the erosion of the old. For others, less certain that all creation was beneficent, Forster's claim was cause for concern. The combination of wonder and fear at corals' marvellous potential was the stuff of the sublime.

In the young science of geology at the time, there was a general concern about the degradation of the present continents and a question about whether and how new land might be formed. The ongoing second great age of discovery had introduced the possibility that newer continents were emerging to replace those older centres of civilisation. In his *Modern Geography* (1802), the mapmaker John Pinkerton wrote that 'the isles of Sunda, the Moluccas, and others in the Indian ocean, are gradually enlarging, and may in time, with Australasia and Polynesia, form a vast new continent, while one or other of the ancient continents will be submerged under the ocean' (quoted in Frost 1988, p. 235). Pinkerton, though he was not in general convinced of the acuity of Forster's observations during Cook's second voyage, did believe that corals were capable of enlarging existing land masses. Joseph Banks, who became president of the Royal Society after travelling as the self-supported botanist on Cook's first voyage, wrote 'Who knows but that England may revive in New South Wales when it has sunk in Europe?' (quoted in Frost 1988, p. 235)

For navigators who followed in the wake of Cook and Flinders, the notion of what mere polyps could do shaped the experience of discovering new reefs and islands. Otto von Kotzebue, the Baltic-German commander of a privately-financed Russian voyage of exploration in the Pacific and Arctic, wrote of his amazement as the European discoverer of an atoll in the group now known as the Marshall Islands.

> The spot on which I stood filled me with astonishment, and I adored in silent admiration the omnipotence of God, who had given even to these minute animals the power to construct such a work. My thoughts were confounded when I consider the immense series of years that must elapse, before such an island can rise from the fathomless abyss of the ocean, and become visible on the surface. At a future period they will assume another shape; all the islands will join and form a circular slip of earth, with a pond or lake in the circle; and this form will again change, as these animals continue building, till they reach the surface, and then the water will one day vanish, and only one great island will be visible. It is a strange feeling to walk about on a living island, where all below is actively at work.[6]

Shocked, as Bougainville had been, that human beings had somehow come to live upon ephemeral slips of land atop coral reefs, Kotzebue wondered, 'to what corner of the earth can we penetrate, where human beings are not already to be found? In the remotest regions of the north, amidst mountains

6 Kotzebue 1821 vol. 2, 36. Emphasis added.

of ice, under the burning sun of the equator, nay, even in the middle of the ocean, on islands which have been formed by animals, they are met with!'[7]

This admiration and gratitude for the fact that corals laboured on behalf of humans was shared by many Europeans who had never seen reefs themselves. Among the most remarkable paeans to the providential power of corals was the 1828 poem 'The Pelican Island', by the British poet, author and hymn-writer James Montgomery. In this reverent epic, which drew explicitly from the voyage narratives of Flinders and Basil Hall, coral 'worms' were the almighty's vehicle for bringing forth new land that man could inhabit.

> Each [coral] wrought alone, yet all together wrought,
> Unconscious, not unworthy, instruments,
> By which a hand invisible was rearing
> A new creation in the secret deep.
> [...]
> Atom by atom thus the burthen grew,
> Even like an infant in the womb, till Time
> Deliver'd ocean of that monstrous birth,
> – A coral island. (Montgomery 1828, p. 18)

'Compared with this amazing edifice, / Raised by the weakest creatures in existence', Montgomery asked, 'what are the works of intellectual man?'[8]

MALEVOLENT CORALS

This belief in the potency of corals had a darker side, however. For those who were more concerned with carrying out European maritime ambitions than with the rather long-term question of where civilisations could flourish when the continent of Europe had been reduced to silt, coral activity was an unalloyed menace. This tension between the positive and negative consequences of coral growth was evident as early as the 1790s, when the surveyor John Macdonald concluded a report on the reefs of Sumatra with an imaginative proposal to harness their work. Though he was convinced that coral rock was derived from plants rather than animals, he attested to the facts 'not only of the very rapid growth of coral, but also of the formation of islands from it, as a necessary, and observed, consequence'. Coral

7 Kotzebue 1821 vol. 2, 36. Emphasis added.

8 On a similar trope of admiration for the humble grains of sand that God had endowed with the power to repel the ocean, see Corbin 1994, pp. 26–27.

structures expanded at a high rate because coral grew upon and consolidated sand and stones, and Macdonald foresaw that the islands of the East Indies would 'in the process of time, become continents or insular tracts or spaces of land' as a result of coral activity. He therefore proposed the creation of a sheltering island in front of the exposed anchorages at Madras, where ships frequently came to grief, by planting coral offshore. 'To attempt to effect this, a considerable quantity of coral might be transported from this coast, at no great expence, and sunk, with stones and other substances, in seven, eight, or eleven fathoms of water. In the course, probably, of forty or fifty years, an island might be formed by the growth of this substance.' In response, the sitting president of the Asiatick Society of Bengal, Sir William Jones, warned Macdonald that the inevitable precursor to potentially beneficial coral islands were treacherous reefs. 'The idea of making islands, for the protection of ships at anchor, is very sublime; but it might be feared, that very dangerous reefs of coral would be formed, before an isle could appear above the water.' (MacDonald 1798, pp. 6–15)

The prospect of corals bringing forth new continents was itself cause for alarm in many quarters. The French botanist and zoologist J.V.F. Lamouroux, whose *Histoire des Polypiers* was primarily a systematic work aimed at revising the taxonomies of Cuvier and Lamarck, contemplated the activity of the reef-building corals with wonder that was matched with concern. He was impressed by the scale of the coral structures that must lie beneath the comparatively insignificant portion of reefs visible at the ocean's surface, and he considered the polyps to be winning a contest with the water. 'Every day the naturalist who traverses the coral islands in formerly water-covered equatorial regions is astonished at the grandeur and the perfect conservation of these masses of polyps; it seems', he marvelled, 'that the sea has just surrendered these areas.' How long would this battle continue in the corals' favour? Lamouroux conjectured that they might flourish so successfully throughout the world's equatorial oceans that they would eventually form a girdle of reefs that blocked navigation between the northern and southern hemispheres altogether (Lamouroux 1816, p. lix).

This idea that corals had the potential, for good or ill, to reshape the globe was dealt a blow in the 1820s by the observations of two French naval surgeons, J.R.C. Quoy and Paul Gaimard. Their examination of fossilised banks of coral upraised at Timor, augmented by their study of living reefs, convinced them that corals could live at a maximum depth of only twenty or thirty feet, and were therefore capable of forming nothing more than thin veneers atop other submarine structures. 'In reconsidering these zoophytes with more attention', they argued, 'we will no longer see them filling the basins of the sea, raising islands, augmenting continents, and menacing future generations by forming a solid circle around the equator.' (Quoy and Gaimard 1825, pp. 289–290) Disputing what they considered to be fanciful claims by Forster and others that instinctive corals were responsible for the annular shape of low islands, Quoy and Gaimard proposed that beneath each ring-shaped reef was a submarine volcano whose crater stood in the shallow depths at which corals could colonise its rim. The Frenchmen succeeded in convincing most contemporary students of geology and zoology that coral activity was limited to a narrow band of depth just below sea level. However, even if corals were busily modifying the submarine topology only in the shallowest thirty feet, these were precisely the thirty feet of depth that mattered for safe navigation.

It was therefore despite Quoy and Gaimard's work that the naval surveyors of Europe became obsessed with documenting and explaining the work of corals. Beginning with the 1831 instructions written for Robert FitzRoy, Commander of H.M.S. *Beagle*, the Hydrographic Office of the British Admiralty made the study of reef formation one of the standard orders for surveys headed to the tropics. FitzRoy was directed to use standard hydrographic surveying methods to examine a coral island. In the end it was his passenger, Charles Darwin, who exploited the island survey to offer a new explanation for the formation of low islands and barrier reefs. Darwin claimed that barrier reefs were formed during the sinking of a landmass that had a reef growing on its shore. If the sinking occurred sufficiently slowly, corals would grow quickly enough to maintain the surface of the reef at or near sea level, and the horizontal distance between the reef and the land would grow as the sloped shoreline receded. If the landmass were an island encircled by a reef, and if the island itself eventually sank completely beneath the waves, what remained would be a ring-shaped reef marking the shape of the former island's shoreline. This theory (which is generally accepted as true today) gained Darwin great acclaim after the voyage; he was a celebrated theorist of reef formation before he had jotted his first private note about natural selection (Sponsel, forthcoming).

Before the *Beagle* had even reached any coral islands, however, back in England the Hydrographer of the Admiralty, Francis Beaufort, had continued his efforts to solve the problem of reef growth. He requisitioned a well-boring apparatus for the 1835 voyage of the *Sulphur* in order that the deep structure of a reef could be studied for clues to its formation. When, in 1842, Francis Price Blackwood was sent to survey the Torres Strait at the northern end of the Great Barrier Reef because so many vessels had been lost after becoming 'entangled within the reefs', Beaufort exhorted him, 'Do not hurry over the hidden dangers which lurk *and even grow* in that part of the world'.[9] Beaufort made the reason for such substantial investments of naval time and money clear in his 1846 directions to Owen Stanley upon dispatching him to make a further survey of the Great Barrier. It was not merely that reefs were dangerous, but that coral growth potentially made the act of surveying marine hazards a fruitless endeavour. '[T]here would be much of discouragement attached to such surveys', Beaufort wrote, 'if changes should be constantly & rapidly at work in [coral] seas'. Stanley was encouraged, therefore, to 'direct [his officers'] attention more particularly to the formation and growth of coral reefs'.[10] If it were possible to account for these processes, one could suggest where reefs were likely to spring up and indicate how long existing charts might be relied on before new reefs would render them obsolete.

A natural consequence of the threat posed by coral reefs was the widespread desire among practically-minded men to find a way to destroy them or limit their growth. Beaufort's American counterpart, the Superintendent of the US Coast Survey Alexander Dallas Bache, instructed the Harvard zoologist Louis Agassiz not merely to study the formation of reefs in the Florida Keys, but to determine 'whether the growth of coral reefs can be prevented, or the results remedied, which are so unfavourable to the safety of navigation' (Agassiz 1852, p. 158). Agassiz argued that there was no possibility of checking the spread of reef building corals 'beyond the bounds which nature itself has [already] assigned to their growth', and urged instead that a close knowledge of the natural tendencies of reef formation would allow mankind 'at least to avoid the evil consequences'. For

9 'Proposed orders for Captain Blackwood'. UK Hydrographic Office (UKHO) MB 3: September 1837–May 1842, pp. 409–416. Francis Beaufort to Francis Blackwood, UKHO LB 12, quoted in Goodman 2005, p. 13(emphasis added).

10 Beaufort reported these concerns from Lord Auckland, then First Lord of the Admiralty, to Owen Stanley on his dispatch to survey the perilous Torres Straits. Francis Beaufort to Owen Stanley, 28 November 1846. UKHO LB 14, pp. 214–215. Emphasis added. See also Cock 2003, n. 420.

example, Agassiz declared, in the Floridian case he believed that the reefs would never extend any farther from the shore than they already did because the slope of the sea bottom outside the present-day reef sank 'rapidly to unfathomable depths' and left 'no opportunity for the growth of a new reef' (Agassiz 1852, pp.158–9). In Britain, the geologist Roderick Impey Murchison hailed Agassiz's work in his 1853 presidential address to the Royal Geographical Society.

> Professor Agassiz has successfully shown how all such surveys ought to be made in conjunction with naturalists. For, quite independent of the important additions to natural-history knowledge which are obtained, statesmen as well as hydrographers thus ascertain the causes of increase or decrease of coral reefs, and learn, that whilst no human power can arrest the growth of such reefs, there are channels amidst them which will remain deep for long periods of time, and the outlines of which, when well defined by lighthouses, may be the salvation of much life and property (Murchison 1853, p. cxvi–cxvii).

The best way to deal with reef growth, then, was to manage its unwanted consequences.

CORAL REEFS DISENCHANTED

Through the rest of the nineteenth century, the enchanted view of corals' activities was in the main limited to popular and fiction writing, while in the specialist scientific literature it was generally agreed that the rate of reef growth was much slower than had previously been feared. Gradually what had been concurrent tropes, of corals as providential wonders and as threats to human enterprise, began to decline.

As a result of his participation in the US Exploring Expedition (1838–1842) to the Pacific, the American geologist James Dwight Dana was more familiar with coral reefs than almost any other Western authority. Although he was a devout Christian, he considered it his responsibility to disabuse his readers of the notion that corals were agents of providence. '[T]o the present day', he complained in 1853, 'the subject [of coral growth] is seldom mentioned without the qualifying adjective *mysterious* expressed or understood ... Very many of those who discourse quite learnedly on zoophytes and reefs, imagine that the polyps are mechanical workers, heaping up these piles of rock by their united labors.' Pointing out that the secretions of stony coral were enclosed within the living polyp rather than being their hives or habitations, he added, '[i]t is not more surprising, nor a matter of more difficult comprehension that the polyp should form coral, than that the quadruped should form its bones, or the mollusc its shell.' (Dana 1853, pp. 47–8) Dana also matter-of-factly dismissed the romantic notion that the low tropical islands were a paradise, commenting, 'The coral island in its best condition is but a miserable residence for a man. There is poetry in every feature: but the natives find this a poor substitute for the breadfruit and yams of more favored lands.' (p. 43) In his 1874 book, *Corals and Coral Islands*, which was intended to present the scientific understanding of the subject to a lay audience, Dana appended a similar set of observations with the note that 'It is not, perhaps, within the sphere of science to criticize the poet. Yet ... more error in the same compass could scarcely be found than in the part of Montgomery's "Pelican Island" relating to coral formations ... There is no "toil", no "skill", no "dwelling", no "sepulchre" in the coral plantation any more than in a flower garden.' He concluded 'The poet oversteps his license, and besides degrades his subject, when downright false to nature.' (Dana 1874, p. 19)

Notwithstanding Dana's pleas, wonderful and industrious corals continued to colonise the pages of popular fiction and children's writing, from the South Sea tales of Robert Louis Stevenson to the children's fiction of Robert Ballantyne and the non-fiction of Charles Kingsley. Dana himself had delighted in many of his experiences on coral islands, 'their groves, and beautiful life, above and within the waters', and many other men of science found that they could not resist the visual charm of a reef.[11] The zoologist Alfred Russel Wallace expressed a common sentiment in writing,

> The bottom was absolutely hidden by a continuous series of corals, sponges, actiniae and other marine productions, of magnificent dimensions, varied forms, and brilliant colours ... It was a sight to gaze at for hours, and no description can do justice to its surpassing beauty and interest. For once,

11 Dana 1874, p. 4. He continued, 'Even the beauty of natural objects had, at times, a dark back-ground. When, for example, after a day among the corals, we came, the next morning, upon a group of Fee-Jee savages with human bones to their mouths, finishing off the cannibal feast of the night; and as thoughtless of any impropriety as if the roast were of wild game taken the day before. In fact, so it was.' Along with the beauty of coral islands, he listed among his 'agreeable memories ... beyond all [the others] ... man emerging from the depths of barbarism through christian self-denying, divinely-aided, effort.'

> the reality exceeded the most glowing accounts I had ever read of a coral sea. (Wallace 1869, vol. 1, p. 463)

By the early twentieth century there were at least two sciences of coral reefs. Whereas nineteenth-century students of reefs such as Darwin and Dana had sought to make themselves expert in the whole process of reef formation, from the growth of individual polyps to the local and global-scale geological and chemical phenomena that created the conditions of possibility for reef formation, most reef scientists of the twentieth century were trained in a particular scientific discipline, usually either geology or zoology/biology. Two transitional figures, both of whom had trained as graduate students at Harvard in the 1890s, were the palaeontologist T. Wayland Vaughan and the zoologist Alfred Goldsborough Mayer (he changed his surname to the less-Germanic 'Mayor' in 1918). In 1904, Mayer founded the Carnegie Institution of Washington's marine biological laboratory at the Dry Tortugas coral islands in the Florida Keys. There he oversaw intensive study of the local reefs, which included pioneering experimental and ecological approaches to the understanding of reefs and their development. Among these were Vaughan's experiments on the rate of coral growth, in which he affixed living corals to terracotta slabs and 'planted' them at different points on the reef. The experiments were aimed at determining the rate of coral accretion, which could be measured by retrieving and weighing the slab to determine the mass of new coral affixed to it, and at understanding how the shape of growing coral colonies was determined by their environmental conditions. Mayor also oversaw and participated in early use of a diving helmet to study reefs from below sea level. At American Samoa in 1919, for example, the new technology allowed him to plant living corals down to a depth of 52 feet. He reported to his sponsor at the Carnegie Institution, 'We really entered a new world in so far as scientific study is concerned when we got down on the submarine precipice under the breakers, and many new facts have come to light.'[12]

While Mayor's audience was primarily scientific, a much wider public was introduced to the experience of coming face to face with the submerged parts of a reef in the following decade by museum exhibitions that included submarine photographs taken during Roy Waldo Miner's helmet dives, and

12 Mayor to Woodward, 28 September 1919. Carnegie Institute of Washington RG Marine Biology, box 1, folder 18, 'Tortugas Lab – Mayer's Trips, 1912–1920.' On earlier use of the diving helmet at Tortugas in 1915, see Stephens and Calder 2006, p. 130.

by the writings of William Beebe.[13] Beebe's 1928 book *Beneath Tropic Seas* described his jaunts on and around the reefs at Haiti in language designed to make them seem at once exotic and accessible. He claimed that '[a]s far as the actual experiences are concerned, any reader of this volume may duplicate them. The diving helmet, hose and pump, with which all the research was done, are as inexpensive as they are simple in operation', and to this end he included appendices on the technical aspects and cost of his outfit (Beebe 1928, p. v). In 1930 Beebe and Otis Barton made the inaugural dives in their 'bathysphere' at Bermuda, descending hundreds of feet deeper than the sun's light reached off the reefs of Nonsuch Island (Beebe 1934).

No sensational dive brought as much attention to a coral reef as did the events of July 1946, when the US Navy tested two nuclear devices on a target fleet anchored in the lagoon at Bikini Atoll in the Marshall Islands.[14] Operation Crossroads was the largest peacetime undertaking that had ever been carried out by the US military, employing 42,000 personnel including a corps of scientists who conducted surveys before and after the bomb blasts. The plan of operations – and by extension the expectations for what was knowable by these surveys – was based on the assumption that an atoll should be understood as a discrete unit. In the minds of those who determined how knowledge would be produced by Operation Crossroads, Bikini was not merely the location where the bombs were tested: Bikini Atoll was, among other things, what the bombs were tested on. The extent to which the bomb's effects over time would be distinguished from seasonal changes in the atoll biota depended on the extent to which an atoll could serve as an experimental system. Bikini emerged as more than a local field site; it became a model to be compared against other islands in the Marshalls that were studied as 'control atolls'. Perhaps the most striking example among the myriad atoll-level studies produced by Crossroads research was Marston Sargent and Thomas Austin's path-breaking paper about Rongelap (which is upwind from Bikini), 'Organic Productivity of an Atoll' (Sargent and Austin

13 'Human Exploration under the Sea', Reidy, Kroll and Conway 2007, ch. 7.

14 According to the official report of Operation Crossroads, the requirements for any test site were 'A protected anchorage at least six miles in diameter … a site which was uninhabited, or nearly so … a location at least 300 miles distant from the nearest city … a location within 1000 miles of a B-29 base … freedom from severe cold and violent storms … predictable winds directionally uniform at all altitudes from sea-level to 60,000 feet … predictable water currents of great lateral and vertical dispersion … [and] control by the United States.' Shurcliff 1947, pp. 16–17.

1949). In a much better known paper that became one of the founding documents of ecosystems ecology, Howard and Eugene Odum cited Sargent and Austin's 'ingenious' methods as inspiration for the Odums' study of 'these inherently stable reef communities' (Odum and Odum 1955).

Among the messages to emerge from the post-bomb resurveys of Bikini were official press releases describing the atoll as a benign laboratory for basic scientific research. The release for 24 July 1947 begins:

> Sun-tanned sailors and scientists observed the anniversary of the world's first underwater atomic bomb explosion today by going swimming in the clear blue-green 84° warm waters of Bikini lagoon. They swam from beaches that one year ago were lashed by high and angry waves thrown outward from the explosion point.
>
> When the Baker Day atomic bomb exploded, half a million tons of water boiled up out of Bikini Lagoon in a mile high cylinder and fell back like rain on the CROSSROADS target fleet in what was probably the most impressive man made spectacle ever staged.
>
> One year later the scientists and military personnel now engaged in an intensive six week scientific resurvey of Bikini Atoll can find few visible effects of that blast. Except for the activities of the 700 man Bikini Scientific Resurvey Task Group, Bikini is the same placid palm ringed lagoon on which [the native Bikinians who had been deported before the tests] King Judah and his subjects sailed in outrigger canoes.[15]

The 1946 tests had been politically controversial demonstrations of power after the previous year's bombings of Hiroshima and Nagasaki, and the Operation Plan for the 1947 resurvey suggests that such press releases were part of a considered strategy to rehabilitate the undertaking.[16] Under the heading 'Public Information Plan', this formerly classified document emphasised '[t]he importance of providing a continuing series of newsworthy press releases to the public, through established public information channels during the course of the operation ... Interesting, newsworthy stories from Bikini, that concern the operation, will forestall much press criticism and speculation of a harmful nature.' These were to include '[s]cience stories concerning the study to be made of the Bikini reefs, and the geological

---

15 'Bikini Scientific Resurvey Press Release No. 16', 24 July 1947. L.P. Schultz Papers, Smithsonian Institution Archives (SIA), Box 26, folder 3, 'Bikini Scientific Resurvey, Correspondence, Press Releases, etc.'

16 On the political controversy, see Graybar 1986, Weisgall 1994.

structure of the atolls'.[17] The most dramatic of the geological studies was the core-drilling carried out by a team of Oklahoma oilmen under the supervision of US Geological Survey palaeontologist Harry Ladd. They bored to 2,500 feet below sea level and retrieved samples aimed at testing Darwin's subsidence theory of reef formation (and at checking the previous year's seismic surveys of the deep structure of the reef, which had been carried out primarily in order to assess the physical damage that the bombs might produce) (Sponsel 2009, ch. 5).

CORAL REEFS SINCE 1950

The results of the drilling and the rest of the geological work at Bikini and later at Enewetak were published in a series of 35 US Geological Survey reports, compiled and edited by Ladd, that began to appear in 1954. The introduction to the first volume was written by Roger Revelle, who had helped to coordinate the scientific operations at Bikini before becoming director of the Scripps Institution of Oceanography in 1950. He began his essay by echoing all the tropes familiar from romantic literature on reefs:

> Of all earth's phenomena, coral reefs seem best calculated to excite a sense of wonder. And of all the forms of coral reefs, the atolls have appeared to men of science to be the richest in mystery and the most strange. Rising alone from the empty sea, these ancient structures, growing now slowly, now fast, toppling when the sea retreats and flung up in haste when sea level rises, are like a Gothic cathedral, ever building yet never finished, infinite in detail yet simple and massive in plan. Tiny plants and animals are their builders. Their architects are the giant ocean and the restless wind.[18]

Upon reading the draft introduction, Ladd's junior colleague Joshua Tracey declared that the 'first two paragraphs are a bit flowery and reminiscent of "The Sea Around Us"', referring to the best-selling 1951 book by Rachel Carson.[19] In fact, Carson scarcely mentioned coral reefs in the first edition of the book. The chapter entitled 'The Birth of an Island' was about Ascension, which is of volcanic origin. Nor was the book particularly oriented toward

17 'Operation Plan' for Bikini Scientific Resurvey, pp. L-1- L-2. (US) National Archives and Records Administration II, RG 374, Defense Nuclear Agency, Entry 47B, Bikini Resurvey, Box 156, Folder A3 'Organization and Management'.

18 Roger Revelle, 'Foreword', in Emery et al. 1954, p. iii.

19 Tracey to Ladd, 5 August 1952 (from Guam). Tracey papers, SIA Accession 02-021, box 4, folder 'Ladd 51-52.'

conservationism or to pointing out harmful effects of human activity. On the contrary, a chapter called 'Wealth from the Salt Seas' touted the opportunities for mineral extraction and petroleum drilling. The idea of concern for the health of coral reefs had yet to be conceived by those familiar with reefs at mid-century.

In the middle of the twentieth century, the prevailing assumption remained that reefs were extremely resilient. This was perhaps partly owing to long traditions that held coral reefs to be durable and incontrovertible hazards of the tropical ocean. In *Beneath Tropic Seas* Beebe described the beautiful reef as a harsh and hardy environment:

> The inner side of the reef was level and shallow, wadable at low water … Farther out, there came an abrupt drop to several fathoms, the barrier wall being tall growths of elkhorn coral, down which, with care, we could clamber to the general reef floor. It did not do to attempt this in any but a calm sea, for every helpless toss against the stone-thorned branches revealed the pitifully thin-skinned defence of our bodies … [T]here was a cruel, interlaced, cobweb of sharp-edged ivory overhead, and escape was possible only by slow deliberate choice of passage. As I painfully made my way down nearer the level of the ground corals, I encountered portières of the stinging millepores. When I reached these I unslung the hammer at my back and pounded off the outer layers (Beebe 1928, pp. 138–140).

If anything, the nuclear weapons tests at Bikini served to reinforce this earlier idea of reefs' resilience. From the perspective of those who didn't happen to share the Bikinians' desire to reinhabit the atoll, the remarkable fact was that Bikini and its nonhuman inhabitants seemed to be recovering normally from the physical impact of the human activities there, while admittedly harbouring a higher proportion of radioactive isotopes than beforehand. Revelle summed up these findings with a redemptive statement: 'Bikini Atoll appears, on the whole, to be a healthy, flourishing structure … The most important groups [of organisms] contributing to the atoll are … [the] same groups [that] have been building up the massive foundation of the present reefs at least since mid-Tertiary times.'[20]

The beginning of activist concern for reefs seems to have coincided with the general rise of environmentalism in the 1960s and 1970s. In their history of science and policy on the Great Barrier Reef, James and Margarita Bowen link the origin of local environmental regard for that largest of reef

20 Revelle, 'Foreword', in Emery et al. 1954, p. vii.

systems to a confluence of events in the mid-1960s. Public and scientific anxiety about an infestation of corallivorous predators (*Acanthaster planci*, the crown of thorns starfish) that had denuded the reef coincided with a new awareness of pollution on the reef resulting from a government-supported increase in petroleum prospecting there after World War II, and with worldwide anger about the effects of pollution on wildlife following the *Torrey Canyon* oil spill off the coast of Cornwall in Britain in 1967. Widespread belief that the government would endorse a controversial proposal to mine the reef for limestone galvanised support for a 'Save the Reef' campaign (Bowen and Bowen 2002, ch. 19). Further explanation for the rise of environmental consciousness among reef scientists is adduced by Jan Sapp in his history of the global *Acanthaster* crisis, in which he argues that many of those who entered reef studies in the 1960s did so because of aesthetic or emotional fondness for reefs, with which they were familiar through avenues including Jacques-Yves Cousteau's 1953 *The Silent World* (Sapp 1999, pp. 77–78, 96–100). Many of them presumably wanted to study reefs using self-contained breathing apparatus, as pioneered by Cousteau and his colleagues and developed through the 1950s.[21]

I can offer a further speculation about the possible role of SCUBA not only in reshaping the direct way in which reefs were studied and perceived, but indirectly in helping fundamentally to modify scientific and cultural meanings and associations attached to the notion of 'coral reefs.' I conjecture that because SCUBA allowed scientists in greater numbers to study reefs underwater, it greatly expanded the proportion of scientists who identified the 'reef' particularly with its living component. For eighteenth- and nineteenth-century naturalists and navigators who arrived at oceanic reefs by way of the very ships that were most vulnerable to destruction on a solid shoal, the 'reef' referred to the whole rigid physical structure upon which a ship might founder. As early as the 1820s, when Quoy and Gaimard advanced the claim that reef-building corals could live only in shallow water, it was widely agreed that the 'living' portion of a reef represented only a tiny fraction of its entire mass. Nevertheless, for most of the nineteenth century, the science of coral reefs was aimed at explaining the origin and shape of the overall reef structure, fossil and living. This remained true into the 1950s, even for a palaeontologist like Ladd, who hoped among other things that his drilling at Bikini would settle fundamental questions about

21 On Cousteau's persona as an explorer, see Kroll 2008, ch. 6.

the formation and history of the whole atoll from its oldest and deepest foundation, including the thousands of vertical feet composed of long-dead material. There were controversies, to be sure, from the mid-nineteenth to the mid-twentieth century over whether these questions would more likely be solved by studying coral organisms or geological phenomena, but the primary scientific question about coral *reefs* was to explain the origin and shape of the whole physical structure (Sponsel 2009, ch. 4–5). Other studies of coral organisms simply belonged to invertebrate zoology.

Before SCUBA, with the few exceptions of those who made helmet dives, scientists who studied coral reefs worked almost entirely above the surface of the water. Lead-lines made it possible to plot the submarine profile of a reef and grappling hooks and dredges made it possible to bring up samples of living and dead coral. One could walk atop a reef flat or make a transect of a coral islet. One could study upraised fossil reefs, but the often-lamented inability to work in any serious way underwater served to reinforce the existing tendency to identify the reef with the whole submarine structure.

It was likely only the post-war creation of a truly underwater science of coral reefs, which, it should be noted, was still limited to the relatively shallow depths at which reef-building organisms flourish, that helped to ensure that a new generation of scientists would find the living portion of the reef to be most appealing and most intellectually intriguing. That so many more scientists and enthusiasts were going underwater also increased the likelihood that events such as episodes of coral disease and transient infestations of coral predators would be noticed and deemed worthy of concern, research and eventual management. While there is now a wide public discourse about the 'death' of coral reefs, little comment is made about the fact that the great bulk of every reef is already dead.

Management has become a virtually ubiquitous watchword for contemporary reef studies, while current knowledge tends to reinforce rather than to resolve the paradoxes presented by coral reefs. As Charles Birkeland writes in the introduction to his edited volume on the science applicable to reef management, entitled *Life and Death of Coral Reefs*,

> Despite having the power to create the most massive structures in the world made by living creatures (including humans), the thin film of living tissue of coral reef seems particularly vulnerable to natural disturbances and the effects of human activities. Coral reefs and other animal-plant reefs have

> always been the first to go during periods of climate change, but they have always come back (although not always with the same components). This combination of attributes – creative power and fragility, resilience and sensitivity, productivity and vulnerability to overexploitation – makes management of coral-reef systems a special challenge to science (Birkeland 1997b, p. 2).

Developments in the sciences of reef ecology and paleoecology have undermined the Odum brothers' claim, made just sixty years ago, that reefs are inherently stable communities. Under the present view that reef ecosystems are fundamentally dynamic, moving between various states of imbalance, complexity is seen to go hand-in-hand with vulnerability. However, as Birkeland's comment indicates, reefs are now recognised to have been extraordinarily robust over geological time. Virtually all the world's reefs have 'died' during various mass-extinction events only to be recolonised and reanimated. Yet there is ever more evidence that reef systems are extremely brittle on the human timescale, susceptible to very rapid deterioration. As Pamela Hallock points out, if one takes the long view '[i]t can be argued that humans are simply part of nature; that exploding human populations are naturally generating another episode of mass extinction from which the Earth will recover in 20 or 30 million years' (Hallock 1997, p. 42). Most of Hallock's colleagues, and most of the rest of us, join her in feeling that the geological perspective – i.e., the likely endurance of the physical reef structure regardless of what we do – is small consolation when so many of the earth's 'living' reefs are mortally threatened by increased sediment deposition caused by deforestation and by the many deleterious effects of anthropogenic climate instability.

CONCLUSION

Like the history of ideas about the deep ocean described by Helen Rozwadowski, the sentiments about coral reefs that I have sketched here were shaped by cultures of exploration, science and popular media (Rozwadowksi 2005). However, the deep sea remained an alien environment which, until recently, could only be encountered remotely and whose very existence could be ignored. Coral reefs, on the other hand, could never be overlooked by those who wished merely to cross the tropical ocean and, far from being alien environments, much of their curiosity arose from the fact that they harboured native humans and held the potential to create new lands to be colonised by civilised society.

Environmental and cultural historians of large animal predators offer accounts that parallel the trajectory I have described for coral reefs, from mysterious, powerful, resilient and treacherous to fragile and susceptible to harm from everyday human activity. Wolves, sharks, whales and even gorillas, once feared and vilified, are now widely viewed as threatened beings that should be prized and preserved as part of the natural world. Just as many biologists feel compelled to defend the predators they study, production of scientific knowledge about reefs now very often goes hand-in-hand with activism for their preservation.

It may be possible to extend this comparison between predators and reefs to include other dangerous or extreme environments and locations such as high altitudes and polar latitudes. The phenomenon now known as the 'rainforest', which is often likened ecologically and metaphorically to coral reefs, has traced a similar arc in Western cultural history. When 'the jungle' was largely unfamiliar and Europeans who encountered it were at comparatively high risk, it seemed both threatening and magnificent.[22] Now that rainforests are (like coral reefs) being damaged by human activities ranging from resource extraction to pollution, they are seen as threatened rather than threatening and have become the object of widespread environmentalist concern.

The narrative I have sketched in this paper suggests several reasons why coral reefs can now seem so benign. Technological changes in navigational instruments and watercraft, as well as the development of weather prediction, sonar and air travel, have reduced the risk that reefs actually pose to human activity. The result of nineteenth- and twentieth-century exploration and mapping is that the circumstances of tropical seafaring are now quite different from what they used to be (although, to be sure, there are plenty of modern boats and tankers stranded upon the world's reefs). Finally, there is ample evidence that the number and extent of healthy reefs is in decline and that a variety of human activities on the local and global scales have contributed to the deteriorating situation. All these features would seem to apply equally well to the histories of animal predators: technology made humankind less vulnerable to predation, and changes to human activity and

22 On Western notions of the 'tropical jungle' and the 'rain forest', see Stepan 2001, Enright 2012.

to the environment mean that fewer humans are placed in harm's way, even while the existence of the predators themselves is threatened.

My purpose in this essay has been to explain why it is no coincidence that coral reefs have occupied a place in culture similar to that of the large animals who make (or made) humans their prey. When coral reefs entered the Western cultural consciousness, their status was precisely that of a mysterious and elusive predator, one that lay in waiting to consume humans, one boatload at a time. Their very name records this fact. The word 'reef' simply referred to any shallow-water area that presented a threat to navigation. 'Reefs of coral rock', or 'coral reefs' as we now call them, were by definition navigational hazards made of or by corals. If our ears were still attuned to this general meaning of reef, 'coral reef' would sound just as ominous as 'man-eater' once did. The Western history of oceanic reefs is inextricable from the way they were encountered and, when encountered in wooden-hulled ships in unfamiliar places, they were no less threatening than the real and imagined beasts in an unfamiliar jungle.

BIBLIOGRAPHY

Agassiz, L. 1852. 'Extracts from the Report ... on the Examination of the Florida Reefs, Keys, and Coast', Appendix 10 in *Annual Report of the Superintendent of the Coast Survey ... for 1851*, ed. by Alexander Dallas Bache. Washington. pp. 145–60.

Beaglehole, J.C. 1974. *The Life of Captain James Cook*. London: The Hakluyt Society.

Beebe, W. 1928. *Beneath Tropic Seas; a Record of Diving among the Coral Reefs of Haiti*. New York; London: G.P. Putnam's Sons.

— 1934. *Half Mile Down*. New York: Harcourt, Brace & Co.

Birkeland, C. 1997a. *Life and Death of Coral Reefs*. New York, NY: Chapman & Hall.

— 1997b. 'Introduction', in *Life and Death of Coral Reefs*. New York: Chapman & Hall. pp. 1–12.

Bougainville, L.-A. de. 1772. *A Voyage Round the World*. Trans. by J.R. Forster. London: J. Nourse and T. Davies.

Bougainville, L.-A. de, M. Bideaux and S. Faessel. 2001. *Voyage autour du monde*. Paris: Presses de l'université de Paris-Sorbonne.

Bowen, J. and M. Bowen. 2002. *The Great Barrier Reef: History, Science, Heritage*. Cambridge: Cambridge University Press.

Bryant, D.G. 1998. *Reefs at Risk: A Map-Based Indicator of Threats to the World's Coral Reefs*. Washington, DC: World Resources Institute.

Burke, L. and World Resources Institute. 2012. *Reefs at Risk Revisited in the Coral Triangle*. Washington, DC: World Resources Institute.

Cock, R. 2003. Sir Francis Beaufort and the Co-Ordination of British Scientific Activity, 1829–55. Ph.D. dissertation, University of Cambridge.

Cook, J. 1955. *The Journals of Captain James Cook on His Voyages of Discovery: The Voyage of the Endeavour*. Ed. by J.C. Beaglehole. Vol. 1. Cambridge: Cambridge University Press for the Hakluyt Society.

Corbin, A. 1994. *The Lure of the Sea: The Discovery of the Seaside in the Western World, 1750–1840*. Berkeley: University of California Press..

Dana, James Dwight. 1853. *On Coral Reefs and Islands*. New York: Putnam & Co.

— 1874. *Corals and Coral Islands*. New York: Dodd & Mead.

Emery, K.O, J.I. Tracey, H.S. Ladd and R. Revelle. 1954. *Geology of Bikini and Nearby Atolls*. Washington: U.S. G.P.O.

Enright, K. 2012. *The Maximum of Wilderness: The Jungle in the American Imagination*. Charlottesville: University of Virginia Press.

Flinders, M. 1966. *A Voyage to Terra Australis. 2, Text*. Adelaide: Libraries Board of South Australia.

Forster, J.R. 1778. *Observations Made during a Voyage Round the World, on Physical Geography, Natural History and Ethic Philosophy*. London: G. Robinson.

— 1996. *Observations Made during a Voyage Round the World*. Ed. by N. Thomas, H. Guest and M. Dettelbach. Honolulu: University of Hawai'i Press.

Forster, J.R. and M.E. Hoare. 1982. *The Resolution Journal of Johann Reinhold Forster, 1772–1775*. London: Hakluyt Society.

Frost, A. 1988. 'Australia: The Emergence of a Continent', in *'Terra Australis' to Australia*, ed. by G. Williams and A. Frost. Oxford: Oxford University Press. pp. 209–38.

Goodman, J. 2005. *The Rattlesnake: A Voyage of Discovery to the Coral Sea*. London: Faber and Faber.

Graybar, L.J. 1986. 'The 1946 Atomic Bomb Tests: Atomic Diplomacy or Bureaucratic Infighting?' *The Journal of American History* 72 (4): 888–907.

Hallock, P. 1997. 'Reefs and Reef Limestones in Earth History', in *Life and Death of Coral Reefs*. New York: Chapman & Hall. pp. 13–42.

Hoare, M.E. 1976. *The Tactless Philosopher: Johann Reinhold Forster (1729–98)*. Melbourne: Hawthorn Press.

Kotzebue, O. von. 1821. *A Voyage of Discovery into the South Sea and Beering's Straits for the Purpose of Exploring a North-East Passage, Undertaken in the Years 1815–1818*. 3 vols. London: Longman, Hurst, Rees, Orme & Brown.

Kroll, G. 2008. *America's Ocean Wilderness: A Cultural History of Twentieth-Century Exploration*. Lawrence, Kan.: University Press of Kansas.

Lamouroux, J.V.F. 1816. *Histoire Des Polypiers Coralligènes Flexibles, Vulgairement Nommés Zoophytes*. Caen: F. Poisson.

MacDonald, J. 1798. 'On the Coral of Sumatra', in *Dissertations and Miscellaneous Pieces, Relating to the History and Antiquities, the Arts, Sciences, and Literature of Asia*, Volume the fourth: being the whole of the articles contained in the fourth volume of the asiatic researches. London: Vernor and Hood. pp. 6–15.

Montgomery, J. 1828. *The Pelican Island, and other Poems*. London: Longman, Rees, Orme, Brown, and Green.

Moureau, F. 2004. 'Philosophes et Marins Francais Dans La Mer Du Sud Avant Baudin: L'exemple de Bougainville et de Ses Compagnons'. *Australian Journal of French Studies* **41** (2): 15–32.

Murchison, R.I. 1853. 'Presidential Address to the Royal Geographical Society of London'. *Journal of the Royal Geographical Society of London* **23**: lxii–cxxxvii.

Odum, H.T. and E.P. Odum. 1955. 'Trophic Structure and Productivity of a Windward Coral Reef Community on Eniwetok Atoll'. *Ecological Monographs* **25** (3): 291–320.

Quoy, J.R.C. and P. Gaimard. 1825. 'Mémoire Sur L'accroissement Des Polypes Lithophytes Considéré Géologiquement'. *Annales Des Sciences Naturelles* **6**: 273–90.

Reidy, M.S., G. Kroll, and E.M. Conway. 2007. *Exploration and Science: Social Impact and Interaction*. Santa Barbara, Calif.: ABC-CLIO.

Rozwadowski, H.M. 2005. *Fathoming the Ocean: The Discovery and Exploration of the Deep Sea*. Cambridge, MA: Harvard University Press.

Sapp, J. 1999. *What Is Natural?: Coral Reef Crisis*. New York: Oxford University Press.

Sargent, M.C. and T.S. Austin. 1949. 'Organic Productivity of an Atoll'. *Transactions, American Geophysical Union* **30** (2): 245.

Shurcliff, W.A. 1947. *Bombs at Bikini: The Official Report of Operation Crossroads*. New York: Wise.

Sponsel, A. 2009. Coral Reef Formation and the Sciences of Earth, Life, and Sea, c. 1770–1952. Ph.D. dissertation, Princeton: Princeton University.

Stepan, N. 2001. *Picturing Tropical Nature*. Ithaca, N.Y.: Cornell University Press.

Stephens, L.D. and D.R. Calder. 2006. *Seafaring Scientist: Alfred Goldsborough Mayor, Pioneer in Marine Biology*. Columbia, SC: University of South Carolina Press.

Veron, J.E.N. 1995. *Corals in Space and Time: The Biogeography and Evolution of the Scleractinia*. Ithaca: Comstock/Cornell.

Wallace, A.R. 1869. *The Malay Archipelago: The Land of the Orang-Utan, and the Bird of Paradise. A Narrative of Travel, with Studies of Man and Nature*. 2 vols. London: Macmillan.

Weisgall, J.M. 1994. *Operation Crossroads: The Atomic Tests at Bikini Atoll*. Annapolis, Md.: Naval Institute Press.

~ 11 ~

The Arctic Ocean as an Outdoor-Laboratory: How the First International Polar Year Expanded the Visual Perception of the Arctic

Alexander Kraus

After a long and sad career of 'academic invisibility', historians have now successfully implemented seascapes and oceans into their research agendas (Wigen 2007, p. 1). Islands, seacoasts and the sea have become the centre of profound analysis, studies have shown how the ocean has been socially constructed throughout history, Michelet's groundbreaking book *La Mer* from 1861 has been rediscovered (Gillis 2004, Gillis 2012, Mack 2011, Steinberg 2001, Michelet 2006). The Arctic and the Southern Oceans have, however, been largely excluded from this development, in no small part because, unlike most other oceans, they are partly covered with floating or solid ice.

Like the seascapes of other oceans, the icescapes of the Polar seas have been perceived or constructed as empty – as 'a space not a place' (Mack 2011, p. 16). Even more important, discoverers like James Cook were astonished by these unknown and icy seas, which made it impossible to distinguish between field ice and land (Wilson 2003, pp. 160–3). These new impressions led Cook to face the limits of his own sea perception. As the scientists of the nineteenth and early twentieth centuries set sail to explore the ocean tides, depths, shores and winds, the frozen Polar seas were also part of these expeditions. This was particularly because the true nature of the Polar regions was still unknown in mid-nineteenth century; speculations ranged from theories of the mystic open Polar sea (Robinson 2007, Petermann 1865) to those of a hollow earth with gigantic maelstroms at the Poles (Felsch 2010, pp. 186–8).

One aim of this paper is to enrol the Polar sea into the history of the oceans, precisely because of its slightly different materiality: the instant of time when the abrupt transformation of seawater into ice started could

PLANCHE III. *21 janv. 1883. Le Varna après la seconde pression violente.*

Figure 1. The Dutch expeditionary ship *Varna*, icebound, after the second devastating ice pressure 1883 (Snellen and Ekama 1910).

be the deciding factor for the success or failure of an arctic expedition. As during the nineteenth century countless passages became icebound, with ships often crushed by ice, the adventurers found themselves in a battle with one of the harshest environments on earth. They had to contend with despair – surrounded by fearsome cold and bleakness, feeling hungry and fearing starvation. No matter whether they failed or succeeded, the press glorified them all the same for their commitment to national pride.

The focus in this paper will be a different one: as most of the innumerable travel reports recounted 'gruesome stories of starvation, hypothermia, even cannibalism' (Reidy 2007, p. 2), it may not be surprising that the visual traditions were closely linked to those of heroic scientific research. But at the end of the nineteenth century there was at least one major exception to the production of national heroic adventure tales: the First International Polar Year 1882–3. In 1875 it was Carl Weyprecht (1838–81), a German-born marine-officer in the Austrian-Hungarian navy, who framed a wide-ranging proposal: the almost unexplored polar region should be examined through a concerted and systematic international cooperation. Until that time, he claimed, polar research had lacked nearly everything that could have been called scientific. In his opinion this deficit was due to the lack of 'simultaneous comparative observation'. A dozen concurrent expeditions were to be sent forth to establish circumpolar research stations, 'all equipped with the same scientific instruments and the same instructions to carry out the same observations for the duration of one year' (Weyprecht 1875a, p. 513). Based on Weyprecht's proposal, the First International Polar Year was initiated in August 1881 by the 3rd International Polar Conference in St. Petersburg. Finally, in 1882, scientific expeditions from more than ten countries set sail to explore the circumpolar region.

In doing so, the researchers turned the polar region from a terrible, mysterious but beautiful seascape into a veritable outdoor-laboratory and enlarged the visual perception of the Arctic. The first section of this paper looks at the visual traditions of arctic representation, the second emphasises the changing scientific approaches in comparing Weyprecht's own scientific research practices during the Austro-Hungarian North Pole Expedition with the later proposed programme of the International Polar Year. As throughout the nineteenth century all kinds of sciences, and especially Arctic research, 'were often and in different ways staged, both by scientists as well as the scientific laity and contemporary media' (Samida 2011, p. 15; see also Riffenburgh 1993), the last section will look at how the Arctic was visualised within national closing reports. This specific expansion of arctic visual traditions is flanked by a closer look at the development of *labscapes* in the nineteenth century: doing field research in a growing world of laboratories (Kohler 2002).

ARCTIC VISUAL TRADITIONS

As Dorit Müller has shown recently for the Antarctic, the visual representations of the scientific and/or geographic expeditions up to the early twentieth century always implemented elements of 'adventure, heroism', the arcane and the 'mysterious'. In so doing they tied in with the age-old visual traditions of the Arctic and Antarctic as untamed and wild (Müller 2009, p. 234). The images – especially of the Arctic region – presented to the public in the nineteenth century through vast panoramas, oil paintings and lithographs in newspapers, journals, arctic adventure accounts and memoirs

Figure 2. Edwin Henry Landseer (1802–1873) *Man Proposes God Disposes* (1864). Wikimedia Commons.

of travellers to the north mostly showed the northern region as 'the Arctic Sublime': a 'drearly terrible',[1] fiery frozen landscape with 'a certain amount of mystery' (Potter 2007, p. 159), with majestic but obviously dangerous cliffs of ice, with endless open spaces and 'these monstrous masses of opaque ice'.[2] A presentation incessantly shifting between the antipodes of 'terror and beauty' (Potter 2007, p. 160). While, before the fateful third expedition of John Franklin (1845–8, none of the men survived), Arctic expeditions were regarded as 'a test for men, a test of their physical and mental endurance, a flirtation with death' (Potter 2007, p. 54–5), the image shifted to the fatal side after it, as Russel Potter remarks: 'The Arctic had been deadly before, but never this deadly; men had been defeated by the elements before, but never so utterly.' (Potter 2007, p. 159)

Sir Edwin Landseer's *Man Proposes, God Disposes* (1864) was one of many paintings which shocked the public at the Royal Academy's annual exhibition of 1864. It shows two enormous polar bears amid the remnants of an arctic sailor, one of them gnawing on a banner which can easily be recognised as a Union Jack. Next to the other, the chest of a human skeleton is visible. The Arctic landscape in the background looks bleak; mighty icebergs reveal the hostility of nature. The painting, praised by its critics for its strong impressions, became a public success; just a few critics and naturally the bereaved regarded it as offensive and tasteless. The accusation

1 W.P. Bayley in an Article for the *Art Journal* 1863 about Church's painting *The Icebergs*, quote in Potter 2007, p. 188.

2 Comment in the *Manchester Guardian* on the above mentioned painting of Church, quote in Potter 2007, p. 187.

Figure 3. William Bradford (1823-1892) *Sealers Crushed by Icebergs* (1866). Museum Syndicate.

of bad taste can be explained as follows: the cold mountains of ice, the vivid cruelty of nature and the bones of the sailors were directly linked by the viewers to the fatal Franklin expedition. In 1864, after a long period of uncertainty,[3] the fate of Franklin and his men was well known to the public, so Edwin Landseer satisfied the public interest by showing the tragic but possible end of the national hero and his men. Ingeborg Høvik has argued that it was such a success because it 'offered a closure to the whole Franklin tragedy that corresponded to British nineteenth-century views on heroism and British-ness' (Høvik 2008).

Another painting that played with a catastrophe of an unspecified expedition was William Bradford's *Sealers Crushed by Icebergs* (1866). The American painter was one of the few artists who travelled the far north by himself and made plenty of drawings in the cold. His majestic landscape paintings presented the polar region as sublime nature, in which humans are clearly wrongly placed – or in which humans do not belong. This is most obvious in the above-mentioned painting, where the shipwrecked sailors are merely helpless lost miniatures next to the natural power of icebergs and ice floes. In his later published travelogue of the 1869 expedition to the coast

3 In 1859, fourteen years after Franklin's ships had set sail to discover the last missing link of the Northwest Passage, M'Clintock's expedition had discovered finally the place where the lost adventurers had perished.

Figure 4. Julius von Payer (1841–1915) *Nie zurück!* (1892). Wikimedia Commons.

of Greenland he often describes nature in terms of bright colours, with sceneries wild, strange and magnificent at the same time (Bradford 1873). As his paintings were described by art critics as very close to nature, one can imagine how they shaped the perception of the Arctic region: 'but Mr. Bradford's art does not profess to concern itself with imaginative or poetic qualities, nor does the artist think it right to take liberties with Nature as Turner did, correcting her defects, atoning for her deficiencies, or teaching her a better way. Mr. Bradford simply records'.[4]

Julius Payer's painting *Nie zurück!* (1892), to give an example of the European continent, is also based on local knowledge and shows the nearly tragic moment as his co-leader Weyprecht, holding a bible in his hand, tries to convince his crew of the Austro-Hungarian North Pole Expedition, after two long winters of privation, to leave the frozen ship behind and walk back towards navigable waters with the sledges carrying the lifeboats. The captured moment is full of drama, emotion and pathos; there is not a single hint on the painting that the crew initially set sail to do scientific research.

4 'Paintings by William Bradford'. *New-York Tribune*, 18 April 1880: 7.

SCIENTIFIC TRADITIONS AND THE EMERGENCE OF THE INTERNATIONAL POLAR YEAR

Next to the adventurous side of Arctic journeys there has always been a scientific side as well. The most important reason among many for travelling the Arctic sea was to find the Geographic North Pole, a chapter of history well known as 'the Race to the North Pole', which began in the middle of the nineteenth century. Carl Weyprecht, introduced above, was one of many who questioned this pursuit, even though he himself had taken part in this race with the Austrian-Hungarian North Pole-expedition. In a talk in 1876, he argued or even railed against every endeavour that had been made towards reaching the North Pole, arguing that this race to the Pole had become a dangerous addiction: in his eyes 'simple' territorial discovery and mapping topography had passed their time as leading sciences as had the traditional approach to polar research. Just trying to reach higher latitudes contributed nothing to the advance of arctic research but only increased personal or national fame. Thus he finally stated: 'Everywhere the people discuss the best route to the Pole, but no-one asks about the scientific treasures alongside these routes' (Weyprecht 1876a, p. 33).

Undoubtedly, scientific research has played a significant role in garnering financial and political support for Arctic campaigns; at least science gave the Arctic explorers 'credibility as men of character', which was considered by Michael Robinson as 'the lifeblood' of expeditions to the polar regions (Robinson 2006, p. 4). Scientific research during the expeditions was needed alongside exploration to help establish 'social authority' at home, even though results and final outcomes barely made their way into public or scientific discussions. Simple scientific and moral rhetoric was always of high importance as it 'functioned as explorers' most powerful tools', but in the end it has always remained '*Stories*, more than specimens or scientific observations, [that] constituted the real currency of Arctic exploration' (Robinson 2006, p. 6).

How meagre the scientific results of cost-intensive polar-expeditions could be becomes apparent by reassessing Weyprecht's own reports and those of his co-leader and later painter Julius Payer (Kraus 2014). The latter can be seen as an example of the classical type of polar-researcher: the heroic masculine adventurer. In his report, written in September 1874 on board the steamboat *Finmarken*, which brought him and his crew back from an essentially failed expedition to discover the North-Eastern Passage and to reach the North Pole, it is clear from the beginning that he sees himself as explorer: as their boat got stuck in the frozen sea earlier than expected, they considered themselves 'not explorers any more but [simple] passengers on a floe' (Payer 1874, p. 14). That the boat drifted towards a previously undiscovered island was both, 'an event of great surprise and importance for the expedition' (Payer 1874, p. 16). Scientific measurement series appear only rarely in his report and if so mostly with a laconic tone: a technical instrument which was brought to practice spectral-analytical observations was simply too weak to bear the intensity of the aurora borealis (Payer 1874, p. 17); the sine qua non of a polar-expedition was the rare chance of a discovery. Payer therefore undertook a series of sledge-expeditions with various companions to discover the new-found land, which brought him fame and glory after his return. Finally, the most touching moment of the expedition arose, 'as for the first time the flag of Austria-Hungaria was blowing in the far north' (Payer 1874, p. 24).

Weyprecht's explanations appear much more prosaic. He describes the discovery of the previously unknown land as only a side note (Weyprecht 1874, p. 6) and focuses instead on the difficult conditions under which a few of them tried to continue the scientific measurement series – mostly of a meteorological and astronomical nature (Weyprecht 1874, p. 5, 7). In his closing comments, he points out that the results of this expedition are of great scientific importance as they provide data from a previously unknown region. His detailed report on the *Scientific Work of the Second Austro-Hungarian Polar Expedition, 1872–4*, published one year later, differs strongly from the first one: after extensive reports on the specific meaning of the first-ever serial magnetic measurements, 'the first regular synchronous observations of the three constants observed in the Arctic region' as he proudly remarks (Weyprecht 1875b, p. 26), he comes to a sobering summary:

> But whatever interest all these observations may possess, they do not possess that scientific value, even supported by a long column of figures, which under other circumstances might have been the case. They only furnish us with a picture of the extreme *effects* of the forces of Nature in the Arctic regions, but leave us completely in the dark with respect to their *causes*. As regards the latter we are as much in the dark as before, simply because we have no synchronous records that might furnish a clue to them ... But as long as Polar Expeditions are looked upon merely as a sort of international steeple-chase, which is primarily to confer honour upon this flag or the other, and their main object is to exceed by a few miles the latitude reached by a predecessor,

those mysteries will remain unsolved. (Weyprecht 1875b, p. 32–3)

Here, as elsewhere (Weyprecht 1875b, p. 25–6), the idea of the International Polar Year shimmers clearly through. To name just one of the above mentioned 'scientific treasures alongside these routes' towards the North Pole, to Weyprecht it was obvious that the solutions to the key questions of meteorology were to be found in the polar regions: 'In the ice of the high north and south are shrouded in secrecy the most interesting questions of atmospheric phenomena, such as the climatology of our globe, and wait to be solved.' (Weyprecht 1876b, p. 9–10) The pole itself, a simple geographic spot, was completely irrelevant for science (Weyprecht 1876b, p. 19). Consequently, he argued that most, if not all, hitherto existing polar expeditions had been following flawed principles, principles opposed to true science and acting in most cases to impede scientific research (Weyprecht 1876a, p. 33). It was therefore the goal to end the race for a simple geographic landmark and at the same time to rediscover the Arctic Ocean as a real and open field of research: observations should include meteorology, geomagnetism, light phenomena like the aurora borealis, ocean currents and tides, water-temperature and many others (Weyprecht 1877).

However, as early as in 1983, Roger Barry remarked that, rather than the different meteorological and earth magnetic findings being seen as the relevant results of this international scientific research project, '[the] main significance lies in the stimulus they provided for the development of suitable instrumentation and observing methods and for greater international cooperation' (Barry 1983, p. 497).[5] Although the nineteenth century had seen precursors in international cooperation (Rothenberg 2009; Rothenberg 2010) – especially in the field of meteorology – the International Polar Year can be and has been interpreted as a 'paradigm shift' in polar research (Krause 2008, p. 384). At the same time, it was a shift which happened without immediate effect: Arctic research after the International Polar Year soon drifted back to former goals; the race to the North Pole continued.

Nonetheless, for a period of time this new research agenda was successfully promoted; American, British and German newspapers and magazines informed their readers constantly about the progress of the undertaking (Kraus 2013). In comparing the work of the International Polar Year with

5 After the death of Weyprecht, the organisation of the International Polar Year lay in the hands of the International Meteorological Organisation (IMO). Ott 1973; Cannegieter 1963, p. 10–18; Baker 1982; Bretterbauer 1985.

previous scientific programmes, widely considered insufficient, journalists sometimes mocked the arctic explorers, claiming that they knew 'neither what to look for nor where to look for it'. As their principal goal was to keep themselves alive, one could wonder whether it is 'scarcely worth while to go so far from home'.[6] After the successful return of the first expeditions, international cooperation was construed as the 'arctic research of the future' (Loewenberg 1886). So how did the scientists of the International Polar Year transform people's imagination of the polar region?

TRANSFORMING THE ARCTIC IN AN OUTDOOR-LABORATORY

'We have become accustomed to the idea', wrote David Livingstone in his concise overview on *Houses of Experiment*, 'that scientific endeavor takes place in specialised locations like the laboratory'. (Livingstone 2003, p. 21) These laboratories quickly became exclusive and restricted places (Kuklick and Kohler 1996, p. 4). But as Robert Kohler has pointed out recently in his reflections on laboratory history, 'placelessness' was one of the hallmarks of the laboratory in the second half of the nineteenth century: 'lab-made facts [were marked] as true not just to their local makers but to everyone, anywhere' (Kohler 2008, p. 766). Ergo the lab itself, as a social form, was able to travel and it was 'easy to adopt' – and easy to translocate. The actual meaning conveyed by the term 'laboratory' has changed several times during the last two hundred years (Gooday 2008, p. 788).

During the time of the International Polar Year, there were, according to Gooday, two major interpretations of laboratory: an indoor research facility or the outdoor field of study. According to the first, 'the naturalist in the laboratory equipped with a microscope learned more that way than by "merely looking at external beauties"', while the second claimed 'that the only legitimate claims to knowledge of organic subjects were those based on authentically outdoor study, not work done in the distorting environment of the institutional laboratory' (Gooday 2008, pp. 789–90). It seems that the scientists of the International Polar Year tried to unite these two opposing positions: they tried to take the lab with all the needed technical instruments out into their arctic outdoor-laboratory. What Kohler has shown for working out in the fields can be adopted to the practice of meteorological science:

6 'American explorers in polar seas', in *The Sun*, 13 April 1884: 4.

'They lived in a world of laboratories, in which they felt bound to use lab methods and understood that their own practices and achievements would be judged by lab standards' (Kohler 2002, p. 4). Other scientists too were faced with similar challenges: in a lot of different sciences, researchers tried to achieve new insights by leaving their now already traditional research area. The scientists of the *HMS Challenger* transplanted the laboratory into the bowels of the ship (Corfield 2001, p. 12); meteorologists constructed 'flying laboratories' in the Age of 'Balloonacy' (Tucker 1996); others wore their scientific equipment on mountain tops looking for a solution of the *mal des montagnes* on site (Felsch 2007, p. 74); while others climbed onto mountains to reach their research-objects, as in the Science of glacial motion (Hevly 1996).

In contrast to the afore-described visual traditions of the Arctic, the vast majority of subsequently published research results of the First International Polar Year were accompanied by pictures of the Arctic Ocean, of coastal environments or of icebergs – yet none of them displayed nature as cruel, brutal, evil, malignant or repellent as had been the case throughout the century before, and as it would remain for at least two or three decades afterwards. The pictorial programme of the International Polar Year did not show the stereotypical battle of man against nature but presented the Arctic Ocean and the littoral as tidy, clean and almost tamed by scientific access.[7] Icebergs and floes still found their way into the images but were presented either as graceful or accompanied by the scientists fascinated by their form, movement, drift and density. Furthermore, every national report of an International Polar Expedition began with a panoramic view

7 See, for example, the pictures in the following volumes: *International Polar Expedition. Report on the Proceedings oft he United States Expedition to Lady Franklin Bay by A. W. Greely*, Vol. I., Washington: 1888; *Exploration internationale des régions polaires 1882–83 et 1883–84. Expédition polaire finlandaise, tome II: Magnétisme terrestre*, Helsingfors 1887; *Beobachtungen der russischen Polarstation auf Nowaja Semlja. II. Theil: Meteorologische Beobachtungen, hrsg. V. R. Lenz*, 1886; *Report of the International Polar Expedition to Point Barrow, Alaska [...]*, Washington 1885; *Beobachtungen der russischen Polarstation an der Lenamündung. II. Theil: Meteorologische Beobachtungen, hrsg. V. R. Lenz*, 1886; *Die Internationale Polarforschung 1882–1883. Die Beobachtungsergebnisse der Deutschen Stationen. Band I. Kingua-Fjord und Die Meteorologischen Stationen II. Ordnung in Labrador: Hebron, Okak, Nain, Zoar, Hoffenthal, Rama, sowie Die Magnetischen Observatorien in Breslau und Göttingen. Herausgegeben im Auftrage der Deutschen Polar-Kommission von Prof. Dr. Neumayer und Prof. Dr. Börgen*, Berlin 1886. The images are also reproduced by the dozen in Barr and Lüdecke 2010.

PLANCHE IV. *23 mars 1883. La maison sur la glace pendant l'hiver.*

Figure 5. The Cabin on the Ice of the Dutch Polar Expedition during the winter (Snellen and Ekama 1910).

PLANCHE V. *9 juillet 1883. La maison sur la glace pendant l'été.*

Figure 6. The Cabin on the Ice of the Dutch Polar Expedition during the summer (Snellen and Ekama 1910).

of the scenery where the particular laboratory was built, showing all relevant buildings. This panorama was followed by a detailed sea chart of the coastal scenery, delivering the exact coordinates and so making it possible to locate the different results on a subsequent global map. What followed next was usually a scale map of the environment itself, denoting every minor building or tent of the camp, such as the magnetic observatory or a tent for magnetic devices, accompanied by lithographs of buildings and occasionally floor plans of one or two of them. Finally the scientific instruments themselves were brought into the spotlight. For nearly every picture, there was a detailed written description of the technical instruments, their installation and the individual test series. Through such presentation, scientists of the International Polar Year not only rediscovered the arctic region as a true area of science but in addition participated in a transformation process which led to a new understanding of field-sciences.

As the Dutch Polar Expedition under the command of Dr Maurits Snellen (1840–1907) became icebound far away from its destination – Dickson Harbour on an island at the mouth of the Yenisey River in northern Siberia – amid the Kara Sea, the scientists decided to erect a prefabricated house on the ice and tried to conduct as much of the proposed research as possible (see Figures 5 and 6). Even though they could not fulfil their task completely, they collected, among other things, extensive information about the meteorology of the Kara Sea, and studied the aurora borealis (Hacquebord 2010; Barr 2008, pp. 223–54). Despite the fact that expedition members witnessed the loss of their ice-damaged ship *Varna,* when the ice started to melt the scientists still continued their work. In doing so, the Arctic itself lost some of its ferocity. The frozen Polar sea had become a place among others to conduct research – a cold but extra large outdoor-laboratory.

BIBLIOGRAPHY

Baker, F.W.G. 1982. 'The First International Polar Year, 1882–1883'. *Polar Record* (Sept.): 1–11.

Barr, W. 2008. *The Expeditions of the First International Polar Year 1882–83.* Calgary: Arctic Institute of North America.

Barry, R.G. 1983. 'Arctic Ocean Ice and Climate: Perspectives on a Century of Polar Research'. *Annals of the Association of American Geographers* 73: 485–501.

Bradford, W. 1873. *The Arctic Regions. Illustrated with Photographs taken on an Art Expedition to Greenland by William Bradford. With descriptive Narrative by the Artist.* London: Sampson Low, Marston Low, and Searle.

Bretterbauer, K. 1985. 'J. Payer, C. Weyprecht, H. Wilczek: The Promoters of International Polar Research', in Wilfried Schröder (ed.) *Historical Events and People in Geosciences.* Frankfurt/M., Bern and New York: Peter Lang. pp. 40–57.

Cannegieter, H.G. 1963. *The History of the International Meteorological Organization 1872–1951 (Annalen der Meteorologie, Neue Folge 1).* Offenbach: Dt. Wetterdienst.

Corfield, R. 2001. *The Silent Landscape. The Scientific Voyage of HMS Challenger.* Washington, D.C.: Joseph Henry Press.

Felsch, P. 2007. *Laborlandschaften. Physiologische Alpenreisen im 19. Jahrhundert.* Göttingen: Wallenstein.

— 2010. *Wie August Petermann den Nordpol erfand.* München: Luchterhand.

Gillis, J. 2004. *Islands of the Mind. How the Human Imagination Created the Atlantic World.* New York: Palgrave Macmillan.

— 2012. *The Human Shore. Seacoasts in History.* Chicago: University of Chicago Press.

Gooday, G. 2008. 'Placing or Replacing the Laboratory in the History of Science?' *Isis* **4**: 783–95.

Hacquebord, L. 2010. 'The Netherlands – Beset in the Ice of the Kara Sea', in S. Barr and C. Lüdecke (eds) *The History of the International Polar Years (IPYs).* Berlin and Heidelberg: Springer. pp. 63–71.

Hevly, B. 1996. 'The Heroic Science of Glacier Motion'. *Osiris* **11**: 66–86.

Høvik, I. 2008. 'Heroism and Imperialism in the Arctic: Edwin Landseer's *Man Proposes – God Disposes*'. *Nordlit* **23**: 183–94.

Kohler, R.E. 2002. *Landscapes and Labscapes. Exploring the Lab-Field Border in Biology.* Chicago and London: University of Chicago Press.

— 2008. 'Lab History: Reflections'. *Isis* **4**: 761–8.

Kraus, A. 2013. 'Scientists and Heroes: International Arctic Cooperation at the End of the Nineteenth Century'. *New Global Studies* **2**: 101–16.

— 2014. 'Scheitern als Chance. Auch eine Vorgeschichte des Ersten Internationalen Polarjahres 1882/83', in M. Aust and J. Obertreis (eds) *Osteuropäische Geschichte und Globalgeschichte.* Stuttgart: Franz Steiner. pp. 175–94.

Krause, R.A. 2008. 'Meteorologie und Geomagnetik als Auslöser der internationalen Polarforschung. Anmerkungen zur Ideengeschichte der Polarjahre'. *Deutsches Schiffahrtsarchiv. Wissenschaftliches Jahrbuch des Deutschen Schiffahrtsmuseums* **31**: 378–96.

Kuklick, H. and R.E. Kohler. 1996. 'Introduction'. *Osiris* **11**: 1–14.

Livingstone, D.N. 2003. *Putting Science in its Place. Geographies of Scientific Knowledge.* Chicago and London: University of Chicago Press.

Loewenberg, F. 1886. 'Hundertjährige Wandlungen in den Zielen der Polarforschung. Zur Säkularfeier John Franklin's, geb. 16. April 1786'. *Die Gartenlaube* **16**: 291–2.

Mack, J. 2011. *The Sea. A Cultural History.* London: Reaktion Books.

Michelet J. 2006. *Das Meer* [1861]. Frankfurt/Main: Campus Verlag.

Müller, D. 2009. 'Fotografie und Südpolarforschung um 1900', in B. Hüppauf and P. Weingart (eds) *Frosch und Frankenstein. Bilder als Medium der Popularisierung von Wissenschaft.* Bielefeld: Transcript. pp. 233–53.

Ott, E. 1973. *Die Weltorganisation für Meteorologie (WMO).* Berlin: Duncker und Humblot.

Payer, J. 1874. 'II', in *Oesterreichisch-ungarische Nordpol-Expedition 1872–1874: Berichte des Herrn Carl Weyprecht und Julias Payer an das Comité der österr[eichisch]-ungarischen Nordpol-Expedition.* Wien. pp. 12–27.

Petermann, A. 1865. 'Der Nordpol und Südpol, die Wichtigkeit ihrer Erforschung in geographischer und kulturhistorischer Beziehung'. *Mittheilungen aus Justus Perthes' Geographischer Anstalt* 11: 146–60.

Potter, R.A. 2007. *Arctic Spectacles. The Frozen North in Visual Culture, 1818–1875.* Seattle and London: University of Washington Press.

Reidy, M. 2007. 'Introduction', in K.R. Benson and H.M. Rozwadowski (eds) *Extremes. Oceanography's Adventures at the Poles.* Sagamore Beach, MA: Science History Publications/USA. pp. 1–14.

Riffenburgh, B. 1993. *The Myth of the Explorer. The Press, Sensationalism, and Geographical Discovery.* London and New York: Belhaven Press.

Robinson, M.F. 2006. *The Coldest Crucible. Arctic Exploration and American Culture.* Chicago and London: University of Chicago Press.

— 2007. 'Reconsidering the Theory of the Open Polar Sea', in K.R. Benson and H.M. Rozwadowski (eds) *Extremes. Oceanography's Adventures at the Poles.* Sagamore Beach, MA: Science History Publications/USA. pp. 15–29.

Rothenberg, M. 2009. 'Cooperation at the Poles? Placing the First International Polar Year in the Context of Nineteenth-Century Scientific Exploration and Collaboration', in I. Krupnik, M.A. Lang and S.E. Miller (eds) *Smithsonian at the Poles: Contributions to International Polar Year Science.* Washington, D.C.: Smithsonian Institution Scholarly Press. pp. 13–21.

— 2010. 'Making Science Global? Coordinated Enterprises in Nineteenth-century Science', in R.D. Launius, J.R. Fleming and D.H. DeVorking (eds) *Globalizing Polar Science. Reconsidering the International Polar and Geophysical Years.* New York: Palgrave Macmillan. pp. 23–35.

Samida, S. 2011. '"Inszenierte Wissenschaft". Einführung in die Thematik', in S. Samida (ed.) *Inszenierte Wissenschaft. Zur Popularisierung von Wissen im 19. Jahrhundert.* Bielefeld: Transcript. pp. 11–23.

Snellen, M. and H. Ekama. 1910. *Rapport sur l'Expédition Néerlandaise qui a hiverné dans la Mer de Kara en 1882–83.* Utrecht.

Steinberg, P.E. 2001. *The Social Construction of the Ocean.* Cambridge: Cambridge University Press.

Tucker, J. 1996. 'Voyages of Discovery on Oceans of Air: Scientific Observation and the Image of Science in an Age of "Balloonacy"'. *Osiris* 11: 144–76.

Weyprecht, C. 1874. 'I', in *Oesterreichisch-ungarische Nordpol-Expedition 1872–1874: Berichte des Herrn Carl Weyprecht und Julias Payer an das Comité der österr[eichisch]-ungarischen Nordpol-Expedition.* Wien. pp. 3-11.

— 1875a. 'Grundprincipien der arktischen Forschung. Vortrag gehalten vor der 48. Versammlung deutscher Naturforscher und Aerzte in Graz von Schiffslietenant Carl Weyprecht'. *Mittheilungen aus dem Gebiete des Seewesens* 11: 505–14.

— 1875b. 'Scientific Work of the Second Austro-Hungarian Polar Expedition, 1872–4 [Translated from the 'Geographische Mittheilungen', vol. xxi, p. 65]'. *Journal of the Royal Geographical Society of London* 45: 19–33.

— 1876a. 'Vortrag gehalten vor der 48. Versammlung deutscher Naturforscher und Aerzte in Graz von Schiffslieutenant Carl Weyprecht', in C. Weyprecht, *Die Nordpol-Expeditionen der Zukunft und deren sicheres Ergebniß, verglichen mit den bisherigen Forschungen auf dem arktischen Gebiete. Vortrag gehalten von* Carl Weyprecht, *k.k. Schiffslieutenant.* Wien, Pest and Leipzig: Hartleben. pp. 27–40.

— 1876b. *Die Nordpol-Expeditionen der Zukunft und deren sicheres Ergebniß, verglichen mit den bisherigen Forschungen auf dem arktischen Gebiete. Vortrag gehalten von* Carl Weyprecht, *k.k. Schiffslieutenant.* Wien, Pest and Leipzig: Hartleben.

[Weyprecht, C.] 1877. 'Graf Wilczek's und Weyprechts Entwurf des Arbeitsprogrammes einer internationalen Polar expedition'. *Mittheilungen aus dem Gebiete des Seewesens* 11: 497–507.

Wigen, K. 2007. 'Introduction', in J.H. Bentley, R. Bridenthal and K. Wigen (eds) *Seascapes. Maritime Histories, Littoral Cultures, and Transoceanic Exchanges.* Honolulu: University of Hawaii Press. pp. 1–18.

Wilson, E.G. 2003. *The Spiritual History of Ice. Romanticism, Science, and the Imagination.* New York: Palgrave Macmillan.

~ 12 ~

The Sea seen as a Tri-Dimensional Social Space: The Case of the Gulf of St Lawrence

Christian Fleury

INTRODUCTION: GLOBAL NATURE AND COMPLEXITY OF A TRI-DIMENSIONAL SPACE

The images of the ocean that we receive from maps often give a false impression of uniformity. The network of commercial routes structures its human activity as do the availability of fishing stocks, or the search for beauty spots by amateur yachtsmen and those on cruise liners. These characteristics are themselves largely dependent on others, which are defined by geographical conditions such as the availability of natural resources for transportation, the distance to continents, the presence of enhancing or facilitating elements such as islands, or restrictive ones such as reefs or ice fields.

As these functional differences have been progressively worked upon and integrated into human societies, so the ocean itself has been subject to a long and evolving process of legal appropriation, from navigational matters to the exploitation of different energy resources. This development has served to highlight its eminently vertical character with four levels of human intervention being identifiable. The uppermost level – the surface – is a place of travel back and forth where freedom of navigation is a norm only challenged by acts of piracy like the Somalian episode, which spectacularly illustrates the historical permanency of that problem. The second level is that of the waters themselves, no longer the sea surface but the nourishing sea, the one that human societies have struggled to share through a complex body of local, international, bilateral and multilateral agreements. The third level is that of the ocean floor which, like the second, is largely concerned with fishing rights although it also constitutes a sensitive area for actors other than fishermen. These latter are predominantly parties involved in extractive activities or the setting up of installations linked to the exploitation of energy resources, be they fossil based or sustainable. Finally, the bottom level – the underwater basement – is the exclusive domain of petrol and gas deposit exploitation.

Whether it be in relation to the ocean's contribution to feeding an ever-increasing human population, the exploitation of its energy wealth, its role in the climatic system, its major contribution to biodiversity or yet again its strategic function as a travel zone in a context of globalisation, humanity is confronted by a promethean undertaking: that of sharing a continuous, global and tri-dimensional space covering 70.9 per cent of the surface of the planet. From both a historical and contemporary perspective, the Gulf of St Lawrence illustrates particularly well how *territorialisation* of the sea, which by its very nature resists the principle of demarcation, leads inevitably to conflict over appropriation. While typically international, conflict in this case occurs within the complex framework of the Canadian federalism.

Measuring 226,000 km^2, the Gulf of St Lawrence is an intra-national sea. Canada was nevertheless relatively slow to consider it to be its exclusive territory, Newfoundland only joining the Canadian confederation in 1949. Further, the size of the fishing stocks had long attracted fishermen from Europe. Finally, it is also important to appreciate that the Gulf serves as access to the high sea for the Great Lakes region which is in turn an international in character. The conflicts surrounding it have tended to evolve vertically, from the maritime to the marine, even the submarine. There is no longer any dispute over the waters themselves. Likewise, the movement of ships is no longer subject to restriction. However the flotillas of foreign fishing boats have seen their activities increasingly curtailed. Finally, questions are being raised as to the modalities of sharing in the event of hydrocarbons being discovered in the seabed.

PIRATES, CORSAIRS AND MILITARY SHIPS IN THE GULF OF ST LAWRENCE

Nothing is known of the nautical misadventures that American Indians may have endured as early users of this great river outlet, itself a significant means of penetrating the interior of the North American continent. The known history of maritime aggressions started with European occupation. Here, as elsewhere, pirates first triggered a sense of insecurity. The Gulf region lends itself to attacking merchant ships. The islands – Anticosti, Mingan and St.

Paul especially – furnish ideal landmarks, while the numerous havens, coves and bays offer natural refuges for those in pursuit of good fortune. Towards the end of the seventeenth century, the administrator of New France, Jean Talon[1] and governor Frontenac[2] encouraged private initiatives with the aim of thwarting the activities of the Spanish or the English. During the entire length of this second hundred years war between 1688 and 1814 which saw European powers in opposition to each other –among whom France and England were paramount – privateers operating in the St Lawrence maritime region represented the North American face of a practice which could be found at its zenith in other parts of the globe. During the war of Succession in Spain (1701–1714), privateering encouraged by governor Vaudreuil allowed the Canadians to compensate for royal shortages.[3] Simultaneously a method of gaining wealth and a source of annoyance for the enemy, it therefore became vital in helping to feed struggling colonies. The taking of Quebec in 1758 and then of Montreal in 1759 did not, however, bring about the demise of French corsairs. For some, such as Joseph Dugas, business continued after the fall of the New-France (Pothier 1980).

Later, during the American War of Independence, other interested parties came to the Gulf. The corsairs originating from the British colonies maintained a well-established tradition of harassment. During the previous century, British colonials from Salem would come to the Gulf intent on disrupting trade amongst the French. Henceforth it was citizens of the United States, encouraged by the letters of marque and reprisal given out by Congress, which threatened British interests. The final seizures in the Gulf of St Lawrence occurred during the Anglo-American War of 1812–1814 (Gardiner 1998).

Security of the maritime route of St Lawrence was a strategic issue during the two World Wars. Germany was looking to disrupt seagoing traffic leaving North America and destined both to support a beleaguered Great Britain and aid in preparations for landings on the continent, as well as, more generally, to carry men, materials and supplies for the gigantic war effort of the Allies. If, during World War One, German operations were focused essentially on a sector of the Atlantic between Cape Hatteras and

1 National Archives of Canada, MGI, Series C 11A, Vol.2, transcriptions: Quebec, 27 Oct. 1667, Talon to the Minister, p. 513.

2 Report 1928–29 of Pierre-Georges Roy, archivist at La Providence of Quebec: Quebec, 15 Oct. 1698, Frontenac to the Minister.

3 See http://www.corsaires.ca.

Halifax (Hadley and Sarty 1991), the second conflict put the river and the Gulf in a more exposed situation. The menacing figure of the U-Boat was the main factor. From May to October 1942 and then from November 1944 to April 1945, multiple attacks, concentrated for the most part on the Gaspe peninsula, claimed hundreds of victims during a phoney war carried out as close as possible to the North American enemy. These attacks were concealed by the Canadian Authorities who drew a veil of silence over the casualties, thus hoping to avoid a collective fear inevitably reinforced by the elusive character of the enemy (Hadley 1985).

Impediments to navigation ceased with the ending of the War. However disputes over sharing of fishery resources have continued to poison relations between Canada and external stakeholders.

THE LONG PROCESS OF EXCLUSION OF EUROPEAN FLOTILLAS FROM THE GULF.

The fishermen of the Gulf region have had their share of problems with their counterparts from New England. Until the 1860s, the latter increased their incursions, leading to problems in coexistence that were occasionally transformed into outright conflict (Desjardins et al. 1999). However it was essentially fishermen originating from further afield who took advantage of the Gulf's resources and did this over a much longer period.

Historical rights are often mentioned in relation to fishing. In this regard, no example is more conclusive in the long term than the annual migration of tens of thousands of European fishermen – notably of French origin – who crossed the Atlantic over a period of some four centuries (Innis 1940). In the case of the Gulf of St Lawrence, British military hegemony in North America effectively limited their incursions. The treaty of Utrecht (1713) greatly weakened French positions by restricting them to Isle Royale (Cape Breton Island) and Isle Saint-Jean (Prince Edward Island). Then the Treaty of Paris (1763) resulted in the elimination of the last French colonies in Canada. Thereafter France only had the small archipelago of Saint-Pierre-et-Miquelon in the region, with its 242 km^2 being set aside for the benefit of French fishermen. By virtue of impacting on French presence along the shorelines of the Gulf, the treaties of Utrecht and Paris also had the effect of reducing French access to the fishing zones of the region. Insofar as possession of land based premises was essential for the success of the type of fishing practised, it was only the exclusivity granted to French fishermen of a portion of the west coast of Newfoundland – called *Petit Nord* by the French and French Shore by the English – that made commercial fishing possible. In 1904, the access to *Petit Nord* was surrendered in the context of the warming of Franco-British relations, the *Entente Cordiale*. However French fishermen maintained access to the adjacent offshore, a three mile zone.

As has been shown, the race for the division of the ocean turned out to be a memorable phenomenon in the second half of the twentieth century. Canada, a country with the longest coastline in the world and the second largest continental shelf, followed this development closely when it wasn't leading it (Beesley 1978, Johnson and Zacker 1977). This proactive policy had repercussions for European activity in adjacent waters. At first, French fishermen were, relatively speaking, spared negative consequences. In fact, when in 1971 Canada extended its sovereignty over the Gulf of St Lawrence, French rights were upheld. However, the following year they lost access to the three mile band of the ex-French Shore when Canada clarified rights of access to the Gulf in establishing a distinction between fishermen from continental France, whose rights ran out in 1986, and those from Saint-Pierre-et-Miquelon, who were not subject to such a deadline.[4]

These provisions bore the hallmark of a privileged relationship with France, to be explained by the historical continuity of French fishing activity in this region and by the proximity to Saint-Pierre-et-Miquelon. However Canada gave the impression of eroding the advantages attached to this special bond when, in 1986, it gave quotas of 3,000 tons to fishermen of the French archipelago, which were judged to be far too small. The period 1986–1988 was especially tense with some of the usual ingredients for conflict in the fishing world being aggravated. A notable example of this was vessels with politicians aboard engaged in illegal fishing, acts that provoked inspections and imprisonments and led to the closure of Canadian ports to French trawlers.

In 1992, in the light of scientific investigations showing depletion of cod stocks, Canada declared a moratorium in its waters. This decision led *de facto* to the end of access to the Gulf as well as to the Atlantic for French ships. In 1994, agreements were signed at a governmental level destined to ease the negative effects of the decisions taken two years previously. They

4 'Agreement between the government of Canada and the Government of France concerning fisheries relations between the two countries', signed in Ottawa on 27 March 1972.

planned to give French rights for a partial return to fishing in the Canadian EEZ proximate to Saint-Pierre-et-Miquelon and also in the Gulf of St Lawrence, but only for extremely limited quantities both in terms of number of species and of level of quotas.[5]

At the beginning of the twenty-first century, in removing the last of the arrangements with European fishermen that might yet have established an historical argument for fishing rights, Canada virtually settled the question of territorial claims to the Gulf of St Lawrence. Despite its very different terms, this question was to resurface in a double-sided context. On the one hand, interest in the region was ignited by the discovery of hydrocarbons and the possibility of their exploitation and, on the other, there were different assessments by the federal and provincial powers as to the sharing of these resources.

THE INTRA-CANADIAN CONFLICTS OF APPROPRIATION IN THE GULF OF ST LAWRENCE

Quebec is particularly concerned by the possible distribution of hydrocarbon resources in the Gulf of St Lawrence, principally because of the location of the Anticosti and Magdalen Islands. It is therefore not surprising that, as early as 1964, the Lesage government took the initiative of framing an agreement with the other four provinces bordering on the Gulf, namely Prince Edward Island, New Brunswick, Nova Scotia and Newfoundland-and-Labrador. This step expressed a common acknowledgement that ownership of the resources of the seabed belonged to the provincial domain. Inter-provincial and extra-coastal limits were thus predicated on the principle of equidistance.

This position was challenged by the federal government, which did not acknowledge any exclusive jurisdiction to any of the five provinces in the sense that it considers the Gulf to be situated beyond their borders. So in the case of Quebec, this is defined in the Royal Proclamation of 7 October 1763 as a line running from the Saint-Jean River at the Cape des Rosiers, then passing through the western point of Anticosti island[6] (Figure 1) to close off the estuary of the St Lawrence River. In other words, 'On the

5 The agreement awarded 2.6% and 1.5% respectively of the Admissible Level of Catch of cod and squid.

6 http://www.tlfq.ulaval.ca/axl/francophonie/Rbritannique_proclamation1763.htm.

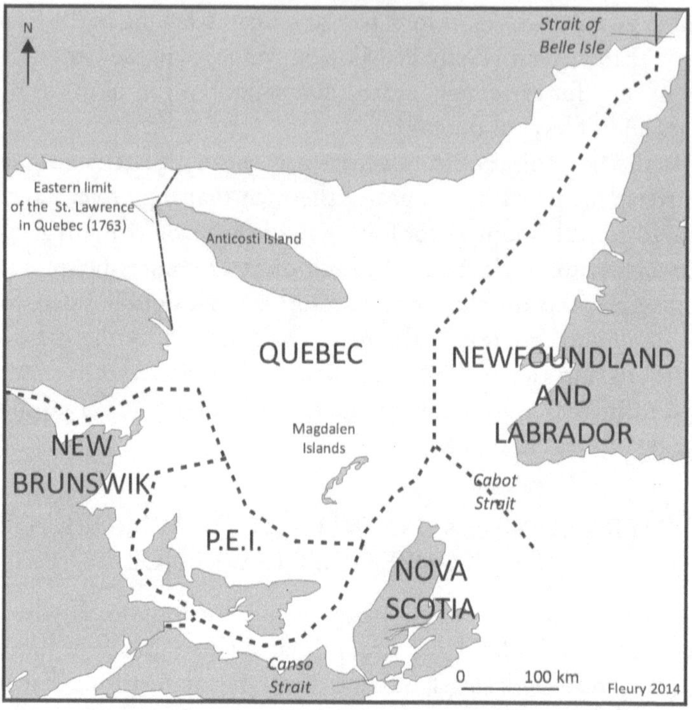

Source : *Ministère des Ressources Naturelles* of Quebec

Figure 1. The Sharing of the Gulf of St Lawrence proposed by the Provinces (1964).

side of the Gulf of St Lawrence, the territory of Quebec ends, in a general fashion, at the low water mark' (Brun and Tremblay 2002).

The question of the provincial jurisdictions in other Canadian maritime sectors was the object of different judgements. Thus, in 1984, a decision at the Supreme Court of Canada relative to Newfoundland-and-Labrador dismissed its claims to the ownership of the rights of exploration and exploitation of the resources of the continental shelf.[7] But this decision immediately led the two parties, both federal and provincial, to conclude an agreement on the distribution of supplies of petrol and gas situated off the Newfoundland coasts without reference being made to any territorial claim.

7 Reference to the Supreme Court of Canada of 28 March 1984 (http://csc.lexum.umontreal.ca).

Nevertheless, another ruling dated 17 May 1982, established a distinction between the case of British Columbia (which entered the Confederation after 1867 – in 1871 to be precise – and which thus had rights over the Straits of Georgia separating Vancouver Island from the mainland) and the bordering provinces of the Gulf.

The line of attack of this question changed in nature in the context of the referendum and possible accession of Quebec to independence. The question of relations between Quebec and Canada viewed as two neighbouring countries had the effect of reawakening interest in the demarcation of a Quebec national territory including a territorial sea, an EEZ and a continental shelf. The study commission on the entirety of the Quebec territory – also called the Dorion commission after the name of its president – opened the question shortly after the deadlock over the disagreement of 1964. This authority reaffirmed the validity of the equidistant method chosen by the interprovincial agreement (Labrecque 1993). In 1991, in the context of the *Law on the process of determining the political and constitutional future of Quebec*, the Study Commission on questions pertaining to the accession of Quebec to sovereignty, commonly named the Belanger-Campeau commission, had also studied this matter and had also confirmed the 1964 line (Charney 1992).

The prospect of an international border in the Gulf of St Lawrence has since receded. However the question of its division – even when considering the federal claim as inalienable – continues to be raised through economic imperatives linked to the very probable presence of important deposits of hydrocarbons in the region. If the principle of provincial sovereignty over the waters of the Gulf is incompatible with Canadian constitutional law, it is still critical for the bordering provinces to determine their respective interests. It is therefore necessary to have boundaries, not necessarily to determine in which province a particular point in the Gulf of St Lawrence is situated, but in order to accurately define the amounts of financial benefits that might accrue to the various territorial bodies, a demand that is necessarily based on a demarcation that is not territorial in the sense of sovereignty but is so in terms of economic benefit.

This was the reasoning behind the decision taken on 26 March 2002 by the court which met in Fredericton and was given the task of resolving the question of demarcations between the provinces of Newfoundland-and-Labrador and Nova Scotia in the Laurentian Channel, starting from the Cabot

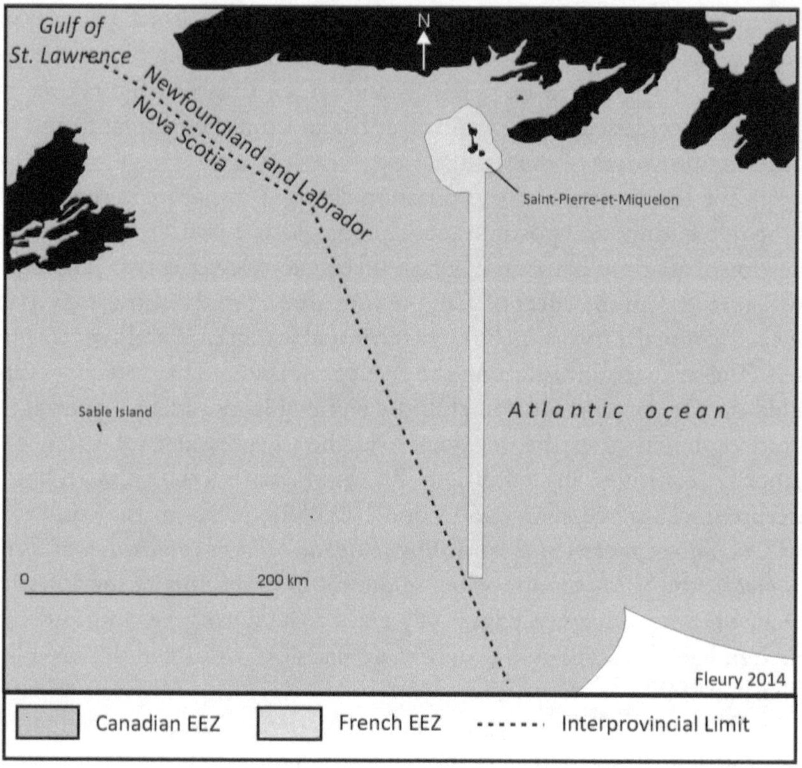

Figure 2. Extra-coastal border between Nova Scotia and Newfoundland-and-Labrador.

Straits and reaching to the outer limit of the Canadian EEZ in the Atlantic. This interprovincial litigation was explicitly linked to the strong probability of important gas deposits being present in a region situated between two *offshore* zones of hydrocarbon extraction in the Canadian Atlantic, Sable Island (Nova Scotia) to the south-west and *Hibernia* (Newfoundland-and-Labrador) to the north-east. The court specified that, if the provinces did not have jurisdictions over the waters in question, they should nevertheless receive a portion of the profits gained from exploitation in federal waters. The logical consequence of the application of the reality principle was therefore to determine an extra coastal demarcation line separating the two provinces,

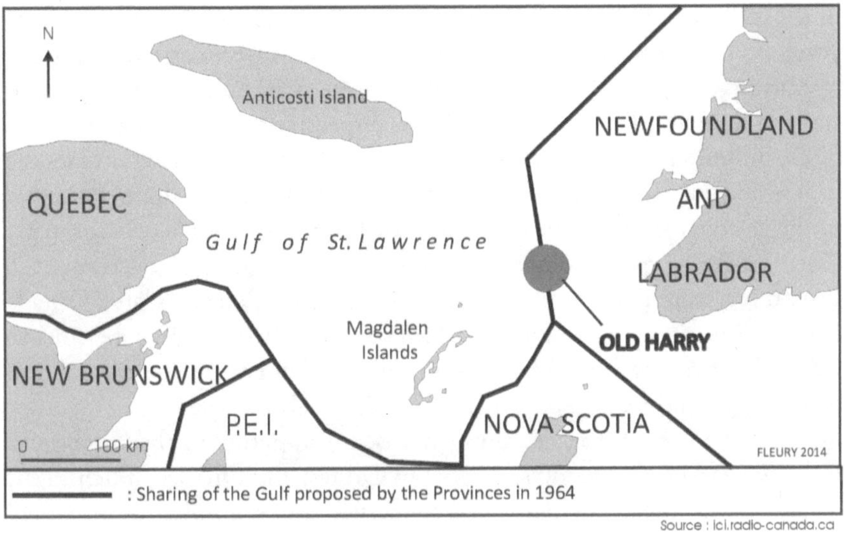

Figure 3. Location of Old Harry.

which was no more and no less than a geodesic line defined by coordinates,[8] exactly as in the case of the drawing up of an international border.

In Quebec, the outcome of this litigation was interpreted as a precedent for provincial aspirations and thus immediately incited reaffirmation of their claims.[9] On this occasion the Quebec Government restated that 'a part of the Gulf of St Lawrence bed belongs to the Province of Québec on the basis of its historical rights to sovereignty and consequently the natural resources present in the sea bed of this territory falls under its jurisdiction'.[10] The reawakening of this question is to be read within the context of the strong probability of the discovery of an important deposit dubbed Old Harry, situated approximately eighty kilometres north-east of the Magdalen islands, overlapping what would be an interprovincial demarcation between Quebec and Newfoundland-and-Labrador (Figure 3). The first operations linked to the hypothesis of a massive presence of hydrocarbons in the seabed

8 *Gazette du Canada*, vol. 137 n° 13 of 18 June 2003.

9 E.g. the declaration of Roger Bertrand, parliamentary representative for Portneuf and Minister for Health and Social Services (http://www.rogerbertrand.org/Communiq/RN030402.htm).

10 Communiqué from the Ministry for Natural Resources and Fauna of Quebec dated 3 April 2002. (http://www.mrnf.gouv.Qc.ca/presse/communiques-detail).

of the Gulf date back to the middle of the 1960s. Over a period of twenty years, no fewer than 80,000 km of seismic surveys were carried out. These operations confirmed the potential of a deposit, without the certainty at this stage of knowing if it was petrol or gas, respectively estimated at 4,000 to 5,000 billion square feet or two to three billion barrels, that is to say twice the importance of Hibernia – where crude oil is extracted – or three times that of the gas deposit of the Ile de Sable.[11] In 1996, the Nova Scotia company Corridor Research Inc. obtained permits from the Quebec government to carry out research in an area containing Old Harry. Two years later, Canada and Quebec decided on a moratorium on the delivery of new permits to search for hydrocarbons whilst awaiting a settlement on the question of ownership of mineral resources in the Gulf's seabed.

If the Fredericton decision whetted the appetites of the Quebecois, it did little to clarify the debate between Ottawa and Quebec, much to the displeasure of the industrial sector concerned and a number of interested politicians. As early as 2002, Hydro-Québec had proposed a partnership to Corridor Resources Inc., and had taken out an option on an initial 25 per cent participation in the drilling programme of the Nova Scotia company, the objective being to obtain a search permit from the Canadian government to drill, as of 2003, the Harry structure 'on the Quebec side' (Hydro-Québec 2002, p. 68). An agreement between the Federal Government of Canada and the Quebec Provincial Government was signed on 24 March 2011 to allow for the development of petroleum resources within the Quebec sector of the Gulf of St Lawrence. However in the absence of any agreement between Quebec and Newfoundland as to the marine boundary between the two provinces – the latter now denounces the 1964 line – matters are at a standstill as of 2014. In addition, to the great displeasure of Quebec, exploratory drilling has been conducted on the Newfoundland side of the boundary from as early as 2010

The possible exploitation of mineral resources in the Gulf of St Lawrence inevitably elevates the question of its appropriation to another level. Indeed, it has led to the intrusion of new parties into the list of those historically involved in conflicts with this marine sector as a backdrop. Different pressure groups – environmentalists, fishermen and residents, to name but a few – have made it known that they are alarmed by the impact of hydrocarbon exploitation in a semi-enclosed sea like the St Lawrence

---

11 http://www.corridor.ca

Gulf. They are mostly part of the St Lawrence Coalition, a broad movement uniting several different associations, including on the Magdalen islands – these being particularly concerned because of their proximity to potential mining installations. In expressing their opposition to the lifting of the 1998 moratorium by the Quebec Government, the various groups stressed the specificity of the Gulf, notably with respect to extended areas of shallow water, the complexity of the flow structure or the severity of the winters. In their view, particular attention should be given to the preservation of such rare or endangered species as blue whale and leatherback turtle, as well as to the long-term future of the fishing activities that continue to generate substantial income for riparian communities. The explosion of the drilling platform *Deepwater Horizon* in the Gulf of Mexico in April 2010 gives them arguments concerning both the eventual consequences for the environment of a similar accident and the shortcomings of compensation guarantees for the affected communities.[12]

CONCLUSION

Throughout the course of history the Gulf of St Lawrence has presented a good example of the range of possible maritime conflicts on, in and under the water. It can thus be considered as a largely socialised maritime space with a number of stakeholders – fishermen, residents, industrialists – operating side by side but each of whose activities can come into competition with each other and even be incompatible. Further, it illustrates a particularly interesting characteristic of internal geopolitics, due to the incidence of two very different problems: the economic pressure to share out the ocean and the complexity of intra-Canadian relations on the federal/provincial and interprovincial levels.

Disputes over ownership point to a specific reality, that of the complexity of the Canadian political system. By virtue of procedures proper to each situation and to each institutional framework, the need frequently arises to manage the economic and financial returns from a marine space that is typically un-territorialised at a sub-national level. Legal arrangements linking coastal territories with segments of marine space will, as such, be required to address projects involving renewable marine energy.

12 www.coalitionsaintlaurent.ca

On a global scale, since 1945 marine space has been the subject of increasing ownership claims by coastal countries. Born from the expectation of laying claim to resources with as yet unknown future economic impacts that are generally expected to be enormous, this major trend in global geopolitics is deployed over an immense space, scientific knowledge of which is considered to be very incomplete, and whose importance in the great natural balance has only been recently understood. This desire to portion out the global ocean calls for corrective mechanisms, which should take two forms. One is universal and leads to the creation of multilateral regulatory bodies to fill the holes or bridge the legal gaps occupied by the criminals and delinquents of the seas. The other is local and seeks to erase the effects of a state-owned territorialisation not adapted to the principal resources of the ocean, whether they be mineral or biological.

If coastal countries progressively utilise legal methods to appropriate the ocean (and they are for the most part in fierce competition, as evidenced by the number of major litigation cases between them), it remains largely up to them to manage this process, not only in relation to their specific needs but also with respect to the maintenance of broad biological and global climatic balances. The sharing of the sea – which can be considered as a functionally stratified space, and moreover one with both fluid and fixed elements – is both urgent for economic reasons and more and more legally complex, as sea-uses become more and more diverse and as the issues become increasingly crucial in nature.

BIBLIOGRAPHY

Beesley, A. 1978. 'Un dossier canadien très chargé : défendre ses droits sur trois mer'. *Fiches d'information sur les pêches*: 39–47.

Brun, H. and G. Tremblay. 2002. *Droit constitutionnel.* Cowansville: Yvon Blais.

Charney, J. 1992. 'Maritime Jurisdiction and the Secession of States: The Case of Quebec'. *Vanderbilt Journal of Transnational Law* 25(3): 343–427.

Desjardins, M. et al. 1999. *Histoire de la Gaspésie.* Montréal: Boréal Express/IQRC.

Dorion, H. 1992. 'Les frontières du Québec: l'état de la question'. *Commission d'étude des questions afférentes à l'accession du Québec à la souveraineté*, pp. 353–374.

Gardiner, R. 1998. *The Naval War of 1812.* London: Chatham Publishing.

Hadley, M. 1985. *U-Boats against Canada. German Submarines in Canadian Waters.* Kingston-Montreal: McGill-Queen's University Press.

Hadley, M. and R. Sarty. 1991. *Tin-Pots and Pirate Ships: Canadian Naval Forces and German Sea Raiders 1880-1918*. Kingston-Montreal: McGill-Queen's University Press.

Innis, H. 1940. *The Cod Fisheries, the History of an International Economy*. Toronto: The Ryerson Press.

Johnson, B. and M. Zacker. 1977. *Canadian Foreign Policy and the Law of the Sea*. Vancouver: UBC Press.

Labrecque, G. 1993. 'La frontière maritime du Québec dans le golfe du Saint-Laurent'. *Cahiers de Géographie du Québec* **33**: 183–218.

Pothier, B. 1980. 'Joseph Dugas'. *Dictionnaire biographique du Canada*. Québec: Presses de l'Université Laval.

Reports

Opportunité énergétique pour le Québec. Corridor Resources Inc.

Plan d'exploration pétrole et gaz naturel au Québec 2002–2010. Hydro-Québec.

~ 13 ~

From Oceans through Islands to Mountains: Creating the 'Correspondence Principle'

Michael S. Reidy

> Like men of every age, we see in Nature what we have been taught to look for, we feel what we have been prepared to feel.
>
> Nicolson (1997)

INTRODUCTION: OCEAN HIGHWAYS

Imagine a large map of the globe, a simple Mercator projection the size of a poster board, with green continents and blue oceans. There is little relief on this map beyond a few shaded areas representing mountain ranges. The ocean is entirely empty. Now, with a red felt-tip pen, begin tracing the early voyages of exploration, one by one, as they embarked from the main ports of Europe. Your red marks will follow one on top of the other, creating a set of ocean highways linked by pivot points in the middle of the oceans. Those pivot points, of course, are islands.

While islands of the past served as stepping-stones for European voyages of exploration, islands of the present serve as the preferred geographies for the study of biogeography. As David Quammen has explained, 'they are natural laboratories of extravagant evolutionary experimentation' where 'a catalogue of quirks and superlatives' present themselves in small, tidy spaces. 'Islands have been especially instructive', Quammen notes, 'because their limited area and their inherent isolation combine to make patterns of evolution stand out starkly' (Quammen 1996, pp. 18–19).

Islands have been studied biogeographically for their isolation, but they are significant historically because of their connectivity. 'Islands are a third kind of place', notes John Gillis, 'something betwixt and between' (Gillis 2004, p. 3). They unite more than they isolate; they connect more than they divide. And the manner in which they connect things – voyages, plants, ideas – is linked directly to their unique physical geography. They are three-dimensional landscapes surrounded by a seemingly endless waterscape. They present to the prepared observer a unique opportunity to revamp the spatial relations of science, the perfect geographies for acquiring a synoptic view.

These two seemingly divergent facets of islands – their geographic isolation and historic connectivity – combined to make them perfect geographies for the foundation of ecology as well. Our environmental consciousness, at least in terms of the devastation wrought by humans on natural ecosystems, began on islands. According to Richard Grove, as early as the sixteenth century, particularly on the Canary Islands and Mauritius, Europeans had already grasped the 'concept of rarity and a fear of extinction' that by the end of the eighteenth century was 'well developed in the minds of French and British colonial botanists' (Grove 1995, p. 9).

In this paper, I will discuss these varied aspects of islands – their isolation, connectedness and ecological significance – as part of a larger question of the process of constructing scientific places. Islands serve as a potent example of the manner in which humans ascribe meaning to a particular geography – transforming a space into a place (Tuan 2001). They are part of a 'collective historical experience', highly constructed physically, symbolically and emotionally through time (Said 200, p. 175).

The first steps on soil outside Europe for virtually every naturalist in the nineteenth century – beginning with Humboldt but including many of his British followers – was on the volcanic soil of islands. Humboldt described with ecstasy the moment he stepped foot on Tenerife, the largest of the Canary Islands, to climb the island's highest peak, the Pico de Teide. 'The ground of the island rises to form an amphitheatre', Humboldt wrote, 'and contains in miniature all the possible climates, from African heat to alpine cold' (Humboldt 1995, p. 34). These same temperature zones, he noted, had been encountered by other naturalists only by travelling horizontally from the tropics to the poles. The correspondence between ocean and island, vividly portrayed as he travelled from sea to peak, led Humboldt to formulate, for the first time, the themes that would occupy him for the rest of his career: the manner in which vertical geographies mirror the horizontal, the interconnectedness of biological and physical forces in nature and, ultimately, his prescient ecological vision, including how humans affect all natural environments.

British naturalists in the nineteenth century followed Humboldt's footsteps (often deliberately up the Pico de Teide). They first stepped from their ocean-bound vessels onto the vertical geographies of volcanic islands. Beginning with Humboldt, the true significance of island geographies lay in the correspondence of these vertical landscapes to the horizontal waterscapes surrounding them. As Kären Wigen has noted, 'the notion of a correspondence between latitude and altitude was in fact a central principle of nineteenth-century geography' (Wigen 2005, p. 25). That 'correspondence principle' became a foundational principle for the study of biogeography, first formulated on islands by naturalists forced to cross the wide expanse of oceans. By following the initial steps of these naturalists, from the oceans to islands and up mountains, I will trace the initial formation of this correspondence principle. In the process, I will integrate a history of ocean-bound voyages of exploration with the terrestrial-focused history of biogeography in order to uncover the preliminary steps toward our modern ecological worldview.

HUMBOLDT AND ISLANDS OF THE IMAGINATION

> No words can evoke the feelings of a naturalist who first steps on soil outside Europe.
>
> Humboldt (1995)

Humboldt's ability to establish an island as a microcosm for the entire globe, as a special, even fortuitous laboratory, enabled him to clearly define the boundaries of biogeography. In essence, he turned islands into controlled, measurable and generalisable environments. Before Humboldt, aspiring naturalists wanted to circumnavigate the globe. After reading Humboldt, they wanted to travel specifically to islands.

To follow the *Pizarro* as it set sail from Europe, follow the thickest red line on your Mercator projection (Walls 2009, p. 51). After the long journey along the ocean's currents, the *Pizarro's* first extended stop was on Tenerife, the largest of the Canary group. Humboldt had finally been transported to a different latitudinal zone, but Tenerife also enabled him to experience the vertical realm as well. The island housed a lofty volcano that extended directly from sea to summit. The *Pizarro's* orders had been set in advance to remain on Tenerife, Humboldt was assured, 'as long as was needed for us to visit the port of Orotavo and climb the Pico de Teide'.[1]

For Humboldt, altitude and latitude came together on this thoroughly explored and increasingly devastated island in the Atlantic. He did not require an untouched wilderness, a virgin peak or a new way of seeing. All he needed was a vantage point, a specific locality that linked horizontal latitudes seamlessly with vast vertical altitudes. That is why he requested that the *Pizarro* stop at Tenerife, and why the first thing he did was climb its highest peak. Even on the summit, he admitted that he saw nothing new. 'What links an excursion to the peak with similar ones to Chamonix or Etna is that one is obliged to follow guides, and sees only what has already been seen and described by previous travellers' (Humboldt 1995 p. 29). What he did see supported a way of viewing nature that he had learned from his former teacher, Carl Ludwig Willdenow, and from the travel narratives of the Forsters, father and son. He saw 'not only the sea to the horizon, but also the forests of Tenerife and the inhabited coastal strips … how plants were distributed according to the decreasing temperatures of altitudes' (p. 35). The experience solidified his vision. He had seen what he was supposed to see and felt what he was supposed to feel. He had hatched his correspondence principle.

He also solidified his ecological vision.[2] Though the term ecological is anachronistic, Humboldt wrote about the devastation wrought by humans in surprisingly modern terms. His lengthy passages concerning the felling of trees and manipulation of water placed humans at the centre of the relationship between the physical and biological realms. Indeed, the main difference between Humboldt's ecological vision and our own is that Humboldt had fully integrated humans into the confluence of forces at work. This confluence, moreover, was for Humboldt a question of social justice, a link we are just beginning to reinvent. He found this on Tenerife as well.

Though Humboldt found what he needed on Tenerife, he chose a different horizontal and vertical space to exemplify his all-encompassing vision. Almost two years later, Humboldt crossed the Cordillera by land to

1 Humboldt 1995 p. 19. The Pico de Teide rises to 3,718 metres.

2 That Humboldt created a surprisingly modern ecological vision with links to social justice is one of the main arguments found in Walls 2009 and Sachs 2006.

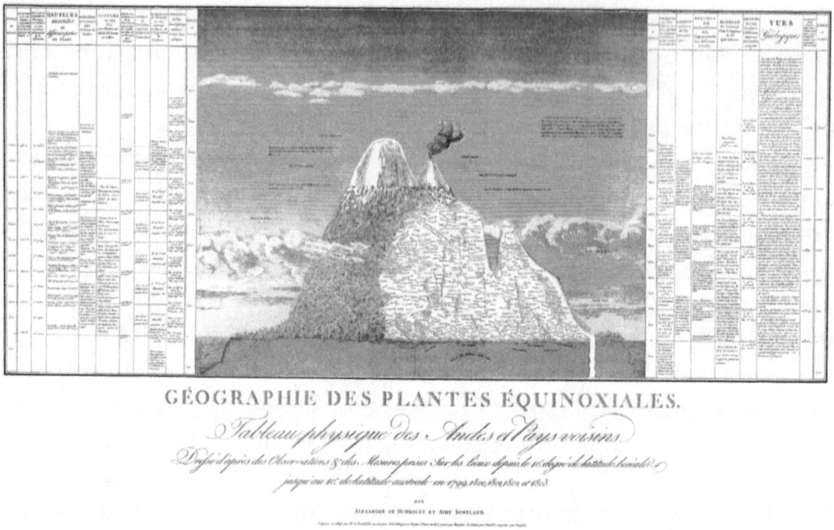

Fiugure 1. Geography of plants in tropical lands. A tableau of the Andes and neighbouring regions. Source : A. de Humboldt, *Essai sur la géographie des plantes* (1805), Vème partie. http://cybergeo.revues.org/25478

Quito, Ecuador, where he attempted to scale Chimborazo, long considered the highest mountain in the world. A thousand feet from the summit, he was forced back by altitude sickness, though he held the world altitude record for over thirty years for this climb. He was extremely proud of his accomplishment. Along with Aimé Bonpland, he had also scoured the countryside for new plants, taken thousands of physical measurements, sketched the topographical and geological features of the Cordillera and interacted with its inhabitants. He was proud of this as well. For the exposition of his correspondence principle, he chose the Cordillera to represent the horizontal and Chimborazo the vertical. The horizontal continent of South America and the vertical topography of its highest peak became his laboratory, the bounded environment he presented as his archetype. It was a sound choice. The vast interior of South America was virtually unexplored, shut off from explorers for more than half a century. Humboldt knew that he could produce the first detailed scientific exploration of the continent. For the vertical realm, the picture he wanted to present was not one of an increasingly devastated island in the Atlantic but, rather, the tallest unclimbed peak in the world, with him gasping for oxygen near its summit.

Humboldt presented this vision to the world in his *Essay on the Geography of Plants*, the first book he published upon his return. It is now considered a seminal text in the history of biogeography (note, however, not island biogeography). In the *Essay*, Humboldt demonstrated how to view the world as a unified connected whole, including physical, biological and cultural elements, all sown together in a three-dimensional synoptic pattern organised around the correspondence of altitude and latitude.

Above all, Humboldt was interested in global patterns, in formulating an overarching synoptic vision. But he had to find a way to do this from particular, specific geographies. The move from the local to the global was fraught with difficulties, problems that Humboldt solved by using a single island to stand for the entire globe, a place where a truly global vision was possible. He used the island of Tenerife as a unifying, archetypal geography that he then transported, first to other islands, then to the larger islands we call continents and finally to the biggest island of all, the globe itself (Nicolson 1995, p. xxv; Jackson, 'Preface' and 'Introduction', in Humboldt and Bonpland 2008, p. viii and 4). In the process, he transformed islands in the imagination of all those naturalists who followed in his footsteps. After Humboldt, islands became the preferred setting for acquiring a synoptic view, and thus the ideal geography to establish a new type of field research. He successfully transformed islands from worlds in miniature to archetypal worlds at large.

Before leaving Tenerife, Humboldt experienced one of his childhood fantasies, to see the Dragon Tree of Orotavo. He then boarded the *Pizarro* and left the island, following the red marks on your Mercator projection. 'We followed the same route', Humboldt explained, 'as Columbus had taken on his first voyage out to the Antilles'. As he watched the island fade into the ocean, with only the summit of the Pico de Teide visible, he deluded himself that he would 'again visit the Canary Islands' (Humboldt 1995, p. 37). He never got that chance; nor did he need to. 'Once discovered', notes Gillis, 'islands remained fixed in place' (Gillis 2004, p. 106). It would be left to other naturalists to visit in his stead.

CHARLES DARWIN AND THE FRUITS OF CORRESPONDENCE

I will never be easy till I see the peak of Teneriffe and the great Dragon tree.

Charles Darwin to Caroline Darwin, 28 April 1831

Little more needs to be written about the role of islands in Darwin's life and work. As the importance of his theory of evolution became rooted in our culture, so too have the Galapagos and island biogeography. The creation of this historical memory surrounding the Galapagos, however, has muted the overarching importance of geographical space, both horizontal and vertical, in Darwin's own thinking. Darwin proffered special status to the study of biogeography in his *Origin*, the only subject that he thought required two chapters. The first of these chapters is dedicated entirely to mountains, the second entirely to oceans. They are arguably the most convincing chapters in his revolutionary text (Sulloway 1979). The correspondence between oceans and mountains, between the horizontal and the vertical, became a unifying vision for Darwin, even before he left England. As the *Beagle* set sail, he was well prepared to see what he was supposed to see.

Like Humboldt before him, Darwin yearned to travel on the currents of the world's oceans. 'In the morning I go and gaze at Palm trees in the hot-house', he wrote to his sister while at Cambridge, 'and come home and read Humboldt'.[3] Above all else, he dreamed of recreating Humboldt's experiences on Tenerife. 'I talk, think, and dream of a scheme I have almost hatched of going to the Canary Islands' (quoted in Harmon 2009, p. 85). When his upcoming voyage was finally in the works, he could hardly check his enthusiasm. 'I hope you continue to fan your Canary ardor', he wrote to his mentor John Stevens Henslow, 'I read & reread Humboldt, do you do the same, & I am sure nothing will prevent us seeing the Great Dragon tree'.[4]

What a sight Tenerife must have been for Darwin as he sat transfixed on the swaying deck of the *Beagle* – and how deep the disappointment never to be able to set foot on its volcanic soil. Quarantined for twelve days offshore, the *Beagle* crew set sail without seeing Dragon Trees or scaling volcanoes. The next set of islands on the route, the Cape Verde Islands,

3 Darwin to Caroline Darwin, [28 April 1831], Darwin Correspondence Project (DCP), Letter 98.

4 Darwin to J.S. Henslow, [11 July 1831], DCP, Letter 102.

served as a disappointing substitute, but the vision he related to the public was the same, one of the land and sea intertwined, with an island forming the nexus. It was from the deck of the *Beagle*, anchored off the coast of the island, that Darwin chose to begin his popular *Voyage of the Beagle*.

> The country rises in successive steps of table land, interspersed with some truncate conical hills, and the horizon is bounded by an irregular chain of more lofty mountains. The scene, as beheld through the hazy atmosphere of this climate, is one of great interest; if, indeed, a person, fresh from the sea, and who has just walked, for the first time, in a grove of cocoa-nut trees, can be judge of any thing but his own happiness (Darwin 1989, p.41).

The vertical viewpoint, described 'fresh from the sea', is repeated throughout Darwin's narrative, his way of mimicking Humboldt's way of seeing. When the *Beagle* anchored off the Falkland Islands, for instance, Darwin related his first impressions through Humboldt's correspondence principle, connecting the distant island to his own. 'Every one has heard of the climate of these regions; it may be compared to that which is experienced at the height of between 1,000 and 2,000 feet, on the mountains of North Wales' (pp. 185–6). When the *Beagle* visited Tahiti, after a 'long passage of 3,200 miles' from the Galapagos, Darwin once again recreated Humboldt's vertical explorations. By ascending the island's mountains, 'sometimes along edges, and sometimes along knife-edged ridges' (Darwin 1989, p. 297), Darwin experienced Humboldt's vegetation zones first-hand. 'The climb up from sea level', explains historian Alistair Sponsel, 'gave him the uncanny feeling that he was crossing great portions of the globe, moving from the equator back toward his home' (Sponsel 2009, p. 104). Darwin had been taught to feel 'uncanny' by Humboldt, to appreciate the connection between latitude and altitude. The need to climb mountains on islands in the middle of oceans dictated both Darwin's thoughts and actions. 'The Beagle staid at St Helena five days', he wrote of his next island visit, 'during which time I lived in the clouds in the centre of the Is[lan]d'.[5]

Darwin's interest in island biogeography heightened after the *Beagle's* return to England. He began a correspondence with Joseph Hooker, then a rising expert in the distribution of plants around the globe. The vast majority of their early correspondence focuses on questions of island biogeography, guided by the correspondence principle. Hooker offered to work through all of Darwin's island material. 'Amongst them', Hooker

5 Darwin to Caroline Darwin, 18 July 1836, DCP, Letter 305.

enthused, 'I hope to find much that is curious and something new too, as it is the alpine plants of the Fuegian Islands which I want more than any thing else'.[6] Darwin replied by pointing specifically to the different elevations, from the 'the summits & bases', in which he had collected plants.[7] 'A discussion on the relations of the Floras', Darwin later instructed Hooker, 'especially the alpine ones, of Azores, Madeira & Canary Is[lan]d would be, I sh[oul]d think, of general interest'.[8] Increasingly, both naturalists were concerned, not with island plants in general, but rather with alpine plants on tropical islands in particular – that is, the most astonishing illustration of the correspondence principle.

After Darwin had come to his theory in the late 1830s, written a pencil sketch in 1842 and a more complete 'Essay' in 1844, the importance he attached to alpine flora on tropical islands diminished. Darwin equated islands with mountains, deserts and other geographies as a means of isolation, what for Darwin had become the 'chief concomitant or cause of the appearance of new forms'.[9] Like Humboldt before him, Darwin was interested in a large-scale, synoptic vision, one that encompassed the entirety of the earth. Although he used examples of island species in his *Origin*, and placed special significance on his time in the Galapagos, his aim was to present a unified vision of zones throughout the earth, latitudinal and altitudinal. It is a connected, integrated worldview, one adopted from Humboldt and transformed to support a comprehensive evolutionary theory.

Like Humboldt, Darwin distorted the manner in which he first came to his views. After verifying Humboldt's correspondence principle on islands throughout the Atlantic and Pacific, by the time the *Origin* was published, his biogeography focused heavily on South America. He chose to begin his revolutionary text with the spatial approach he had gleaned from Humboldt, but the physical geography had changed. 'When on board H.M.S. *Beagle*, as naturalist', Darwin famously began, 'I was much struck with certain fact in the distribution of the inhabitants of South America and in the geological relations of the present to the past inhabitants of that continent' (Darwin 1988, p. 65). As with Humboldt, the need for an archetypal setting prompted this move. Darwin's theory rested on both

6 Hooker to Darwin, 28 November 1843, DCP, Letter 717.

7 Darwin to Hooker, 31 March [1844], DCP, Letter 744.

8 Darwin to Hooker, 25 December [1844], DCP, Letter 803.

9 Darwin to Hooker, [8 September 1844], DCP, Letter 776.

a transformation of space (following Humboldt) and a transformation of time (following Lyell). For temporal, geological purposes, South America worked best. Fossils had been collected there for some time, and this was where Darwin made the correlation between extant species and their fossil ancestors. Darwin transferred his spatial, geographic approach to South America as well. The manner in which Darwin came to his views and the subsequent manner in which he later articulated those views in print, therefore, matched that of Humboldt. Both had initially come to their views on islands, and both carried these initial insights to larger, more archetypal settings on the South American continent.

JOSEPH DALTON HOOKER AND THE PROBLEM OF CORRESPONDENCE

> I was greatly pleased with finding my most Antarctic plant, *Lecanora miniata* at the top of the pass, and today I saw stony hills at 19,000 feet stained wholly orange red with it, exactly as the rocks of Cockburn Island were in 64 degrees South; is not this most curious and interesting? To find the identical plant forming the only vegetation at the two extreme limits of vegetable life is always interesting.
>
> Joseph Hooker to William Hooker, 13 September 1849

Lecanora miniata is a species of lichen that thrives in extreme environments. When Hooker first sighted it on the rocky shores of Cockburn Island off the coast of Antarctica, he was serving as the official naturalist on the famed James Clark Ross expedition to the South Pacific (1839–1842). When he saw it again ten years later, he was on the Sikkim–Tibet border, perched precariously high on the side of a cliff well above 18,000 feet, with streaming glaciers and frozen lakes below. Hooker, it seems, was drawn to the extremes, a trait that he shared with the lichen he was hunting.

Yet, as 'pleasing' and 'interesting' the find was, it was certainly not very 'curious'. Hooker was well prepared to see what he was supposed to see. He had travelled all that way specifically to find just such a connection. He had made his name as a high latitude explorer before he became a mountain traveller. He had written several books on island biogeography before he wrote his book on the Himalayas. All are spatial and Humboldtian, and the important transition from the horizontal to the vertical occurred on islands.

The four-year sojourn of the *Erebus* and *Terror*, with stops at as many of the pivot-points as possible, offered the young Hooker plenty of islands to plunder and ponder. Published in separate instalments upon his return to England, *The Botany of the Antarctic Voyage* is a testament to the vertical and horizontal geographies he had created in his mind from reading Humboldt. The crew's first stop was Madeira, where they remained for eleven days, then Tenerife (Humboldt) and the Cape Verde Islands (Darwin). Three long tours in the Southern Ocean followed, all botanically represented by extended sojourns on islands. In the first tour south, three islands predominated, a two and a half month anchorage at Kerguelen's and extended stays at Lord Auckland's and Campbell's. The second tour entailed a 'prolonged stay' in the Falklands, and the third cruise focused on the visit to Cockburn's Island (where he first found his *Lecanora miniata).*

Though the publication is his, and he did all the travelling, the questions he asked were those of Humboldt and Darwin. He viewed islands through their eyes. Islands not only served as central connecting points, the pivots of his voyage, but also represented the perfect geographies to form a unifying vision of plant distribution. 'Hence it will appear that islands so situated furnish the best materials for a rigid comparison of the effects of geographical position and the various meteorological phaenomena on vegetation, and for acquiring a knowledge of the great laws according to which plants are distributed over the face of the globe' (Hooker 1844, p. xii). These great laws for Hooker, as they had been for Humboldt and Darwin, were synoptic and global. His account of Campbell's Island, for instance, where mountains rise abruptly to almost 400 metres, is a Humboldtian account in miniature, with the correspondence principle providing the means of seeing and feeling. Searching for one plant in particular, *Gentiana*, Hooker was struck by its scarcity and relieved once he found it. 'The fact of the occurrence, and the great number, of species of *Gentiana* inhabiting only the more elevated regions of the temperate and tropical zones, and there reaching the snow limit, renders it very remarkable that they should be so proportionally scarce in the higher latitudes of the northern and southern hemisphere. Generally speaking, the inhabitants of these elevated and cold regions are species of such Natural Orders and genera as compose the mass of the Polar vegetation' (p. 57). It was 'very remarkable' because it did not fit into Humboldt's or Darwin's global vision, and he could speak 'generally' because that was what mattered, even when such a generality appeared unsustainable. He was looking for what he was taught to look for, and once found, he felt what he was supposed to feel.

By thc timc Hooker was working on his publication, problems were beginning to arise in the global patterns of distribution. Concerning species of ferns on St Helena and Ascension, he was struck by the difference of species on islands set close together geographically. These facts were difficult for Hooker to interpret. He had been well prepared to see a unifying vision, yet the overarching law-like nature of plant distribution eluded him (Browne 1983).

Figure 2. Campsite in the Himalayas, from Joseph Hooker, Himalayan Journals, 1854. https://archive.org/details/himalayanjourna00hookgoog

When Hooker was planning his next great exploratory voyage, these same questions were still haunting him. His first great voyage of exploration gave him a thorough grounding in the vegetation of islands, but very little in the way of global, overarching laws. Climbing in the Himalayas, Hooker reasoned, would enable him to explore and compare three distinct vegetation zones in a single day: a tropical zone, a temperate zone and an alpine zone (Reidy 2010, p. 29). His pioneering treks in 1849 and 1850 into the heights of the Himalayas, and his resulting *Himalayan Journal* (1854), constituted a testing ground. Guided by the correspondence principle, he was there to supplement his extreme latitudinal studies with equally extreme altitudinal ones, all with the goal of finding synoptic laws that governed the distribution of species. Although the trip was helpful for Darwin, who subsequently used much of his friend's material in the *Origin*, it failed to answer Hooker's own overarching questions. He was stuck in the rut of the correspondence principle. Perched high on the cliffs above the Tibet–Sikkim border, he found the species of *Lecanora* for which he was searching, but little else.

ALFRED RUSSEL WALLACE AND THE DEMISE OF CORRESPONDENCE

> Mere distance is one of the least important of the causes which have determined the likeness or unlikeness in the animals of different countries
>
> Wallace (1880)

Like Hooker, Alfred Russel Wallace went on two major expeditions that helped establish his career. The first was an extended sojourn through the Amazon. His second trip, a decade-long excursion to the Malay Archipelago (present-day Indonesia and Malaysia), transformed him from a collector into an evolutionist, and from a landlubber into an island-hopper. He became the premier island biogeographer in the nineteenth century. He returned to England in 1862 with enough material to publish numerous books and articles on the distribution of plants and animals, including his classic travel narrative, *The Malay Archipelago* (1869), the two-volume *Geographical Distribution of Animals* (1876) and his seminal *Island Life* (1880).

In these texts, oceans and islands took centre stage. While his first text, *The Malay Archipelago*, was primarily concerned with the animals and plants on each island, and the differing languages and customs of their natives, it did allude to the strict division of animals between the islands of Bali and Lombok, what became known as 'Wallace's Line'. In his next text, *The Geographical Distribution of Animals,* Wallace expanded on this still inchoate concept of divisions between regions, extending it to the grand scale of the earth. It was a break from previous work on the distribution of species, focusing on 'zoological regions' – not altitudinal or latitudinal zones (Wallace 1876, pp. vi–x). The accompanying illustrations are primarily maps depicting each of these regions or subregions. Though he also included a series of twenty plates illustrating animals and plants, they too are organised geographically, emphasising the 'physical aspect and the special zoological character' of each 'well-marked division of a region'. 'Zoological regions' had replaced altitudinal and latitudinal 'zones of vegetation'. The vantage point had changed as well, from the cross-sectional views of Humboldt to a global projection of asymmetrical regions. His frontispiece, for instance, a large Mercator map of the globe – 'The World on Mercator's Projection shewing the ZooGeographical Regions and the Approximate Undulations of the Ocean Bed' – depicts different zoological regions and different depths of the ocean according to number and colour.

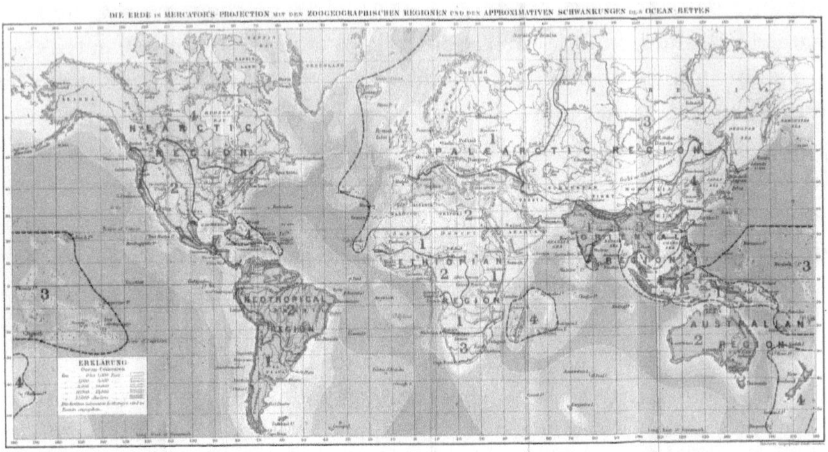

Figure 3. Map of terrestrial zoogeographic regions of the world published in *The Geographic Distributions of Animals* (1876) by Alfred Russell Wallace. http://macroecology.ku.dk/resources/wallace/wallace_map.jpg

Humboldt's influence remained visible in these maps as well. Along with the importance of the ocean depths, Wallace included the 'zones of altitude above the sea level' as also contributing fundamentally to the distribution of animals. His maps of each 'zoological region', therefore, also included 'belts of altitude' depicted through contour shading (pp. ix–x). Thus, horizontal and vertical spaces retained their importance; they were, however, depicted in a different manner, for different reasons and with an entirely different result. There is no correspondence principle in Wallace's new scheme.

Wallace expanded on this theme in *Island Life*, a book that cemented his differences from past researchers. He began on the ocean: 'When an Englishman travels by the nearest sea-route from Great Britain to Northern Japan he passes by countries very unlike his own, both in aspect and natural productions'. Yet, after a 'circuitous voyage of thirteen thousand miles', Wallace continued, 'when he visits the interior of the country he sees so many familiar natural objects that he can hardly help fancying he is close to his home' (Wallace 1880, pp. 3–4). Similar sea-voyages had been undertaken by previous naturalists, their eyes taught to see what they were supposed to see, transfixed on zones of vegetation dictated by latitude and altitude. How could they explain, Wallace asked, the extreme diversity found on islands within a day's journey? 'In the Malay Archipelago there are two islands, named Bali and Lombok, each about as large as Corsica, and separated by a strait only fifteen miles wide at its narrowest part. Yet these islands differ far more from each other in their birds and quadrupeds than do England and Japan' (p 4). No difference in climate, soil or position could account for it. The problem, moreover, appeared universal, exhibited on 'every continent, every country, and every island on the globe' (p. 6). According to Wallace, no universal law of correspondence would help solve it.

Wallace attempted his own solution in *Island Life*. It is a text that uses the physical geography of islands, combined with the mechanism of natural selection, to analyse the complex interaction between physical and biological forces that explains where animals and plants are found and why. Several classes of facts are needed, according to Wallace, including adequate knowledge of the flora and fauna of the world, the geological history of the earth and 'the exact depth and contour of the ocean-bed' (p. 10). This last point was crucial. A distinct and thorough knowledge of the ocean was required to understand both the former existence of submerged lands that united islands to continents and the deepest parts of the ocean that isolated them. It was, for instance, the most significant clue for explaining 'Wallace's Line'. The fifteen-mile strip of ocean between Lombok and Bali is extremely deep, forming the necessary barrier between zoological regions. Here, as elsewhere, the depth of the ocean proved far more important than distance.

As suggested by the title of the text, islands for Wallace were the essential physical geographies for studying the complex confluence of biological and physical forces at work. Their 'restricted area and definite boundaries' present the distribution of species in 'their simplest form' (p. 233). They serve as a methodological tool, the first place to study before expanding out to the larger landmasses of the world. 'When we have mastered the difficulties presented by the peculiarities of island life we shall find it comparatively easy to deal with the more complex and less clearly defined problems of continental distribution' (p. 234). His method was similar to Humboldt's and Darwin's, to begin with islands and transfer the lessons to continents. However, his viewpoint was a radical departure from those before him. He replaced the correspondence principle, which had guided research for over eighty years, with 'areas of distribution' and 'zoological regions' (p. 500). The correspondence principle had run its course.

CONCLUSION

> Doth not a Tenerife, or higher Hill
> Rise so high like a Rocke, that one might thinke
> The floating Moone would shipwrack there, and sinke?
> ...
> Are these but warts, and pock-holes in the face
> Of th' earth?
>
> John Donne, 'An Anatomy of the World' (1611)

The way we approach the natural world is conditioned by our culture. John Donne lived during a time that viewed mountains, such as Tenerife, as abominations, mere warts on an otherwise beautifully round planet. The transition, from mountain gloom to mountain glory, was relatively fast-paced; in a period of fifty years mountains began to elicit grandeur rather than the grotesque (Nicolson 1997).

Humboldt was a product of the Romantic age that had transformed our conceptions of mountains. By the time he sailed on the *Pizarro*, Tenerife was primed to elicit far more than disgust. Humboldt used it to solidify an entirely new conception of space in the production of knowledge. Before the nineteenth century, most naturalists studied the biogeography of plants by encountering different zones of vegetation as they sailed from the higher latitudes to the equator. Zones of latitude drove the study of biogeography. With the work of Willdenow and Forster as his guide, Humboldt transformed this process by transforming the space of islands into a place for science. As archetypal geographies uniting horizontal latitudes with vast vertical altitudes, islands could now serve as microcosms for the entire global environment. Surrounded by oceans and reaching into the heavens, islands could be illustrative and comprehensive, a physical space where a global vision was possible. For Humboldt, nothing was in isolation. Everything was connected. His approach to nature prepared him for this. Tenerife demonstrated it. He connected oceans and mountains through islands.

After Humboldt, naturalists learned to think *with* islands (Gillis 2004, p. 107). They followed in his footsteps, using islands to experience the 'correspondence principle'. On islands, naturalists could symbolically travel around the world, from the equator to the poles, from the ocean to the summit of mountains. For Darwin, Hooker and other British naturalists, the combination of the horizontal with the vertical was central to their thinking as they searched for the unifying laws of biogeography. Humboldt's principle directed the science of biogeography for three quarters of a century. Most naturalists recognised anomalies, often challenging Humboldt's relatively simple views on the relationship between temperature and plant life (Browne 1983). Yet, the principle itself remained firmly entrenched, underpinning how naturalists were supposed to look at the world. It was not questioned because it had become unwittingly assumed.

Wallace was the first researcher to fully recognise the problems in the correspondence principle and offer the most enduring solution. Mere distance, temperature or any such simple correlation would no longer do. The vertical orientation of Wallace's thinking, present even in his chaotic and untidy maps, is a testimony to the enduring influence of Humboldt, even if the visual productions and scientific conclusions are different. The reciprocal influence of oceans and landmasses was retained, as was the fundamental role of islands as stepping-stones for migration and physical geographies of transmission. And lastly – Humboldt would say most importantly – the overarching lessons of Wallace's text, the apt moral to be gleaned from his tale, is Humboldtian to the core. 'I trust that the reader who has followed me throughout', Wallace concludes *Island Life*, 'will be imbued with the conviction that ever presses upon myself, of the complete interdependence of organic and inorganic nature' (Wallace 1880, p. 511). Wallace ends where Humboldt had begun, with how we should approach the natural world, a confluence of the physical and biological, the inorganic earth with organic life. Our term for this is 'ecology'.

BIBLIOGRAPHY

Browne, J. 1983. *The Secular Ark: Studies in the History of Biogeography*. New Haven, Connecticut: Yale University Press.

Darwin, C. 1988. *The Origin of Species*. Ed. by J.W. Burrow. New York: Penguin.

— 1989. *Voyage of the Beagle*. Ed. by J. Browne and M. Neve. New York: Penguin.

Darwin Correspondence Project (www.darwinproject.ac.uk)

Gillis, J.R. 2004. *Islands of the Mind: How the Human Imagination Created the Atlantic World*. New York: Palgrave Macmillan.

Grove, R. 1995. *Green Imperialism: Colonial Expansion, Tropical Island Edens, and the Origins of Environmentalism*. Cambridge: University of Cambridge Press.

Harmon, P.M. 2009. *The Culture of Nature in Britain, 1680–1860*. New Haven, Connecticut: Yale University Press.

Hooker, J.D. 1844. *The Botany of the Antarctic Voyage of HM Discovery Ships* Erebus *and* Terror *in the Years 1839–1843, Under the Command of Captain Sir James Clark Ross*. London.

— 1854. *Himalayan Journals: Or Notes from a Naturalist in Bengal, Sikkim and Nepal Himalayas, Khasia Mountains*. London.

Humboldt, A. von. 1995. *Personal Narrative of a Journey to the Equinoctial Regions of the New Continent*. Ed. by J. Wilson. New York: Penguin.

Humboldt A. von and A. Bonpland. 2008. *Essay on the Geography of Plants*. Ed. by Stephen T. Jackson. Chicago: University of Chicago Press.

Nicolson, M. 1995. 'Historical Introduction', in Alexander von Humboldt. *Personal Narrative of a Journey to the Equinoctial Regions of the New Continent*. Ed. by J. Wilson. New York: Penguin. pp. ix–xxxiv.

Nicolson, M.H. 1997 [1963]. *Mountain Gloom and Mountain Glory: The Development of the Aesthetics of the Infinite*. Seattle, Washington: University of Washington Press.

Quammen, D. 1996. *Song of the Dodo: Island Biogeography in an Age of Extinction*. Simon and Schuster.

Reidy, M.S. 2010. 'From Oceans to Mountains: Constructing Space in the Imperial Mind', in J. Vetter (ed.) *Knowing Global Environments: New Historical Perspectives on the Field Sciences*. New Brunswick: Rutgers University Press. pp. 17–38.

Sachs, A. 2006. *The Humboldt Current: Nineteenth-Century Exploration and the Roots of American Environmentalism.* New York: Viking.

Said, E. 2000. 'Invention, Memory, and Place'. *Critical Inquiry* **26**(2): 175–92.

Sponsel, A.W. 2009. 'Coral Reef Formation and the Sciences of Earth, Life, and Sea, c. 1770–1952'. Ph.D. Dissertation, Yale University.

Sulloway, F. 1979. 'Geographic Isolation in Darwin's Thinking: The Vicissitudes of a Crucial Idea'. *Studies in History of Biology* **3**: 23–65.

Tuan, Y-F. 2001. *Space and Place: The Perspective of Experience.* Minneapolis, MN: University of Minnesota Press.

Walls, L.D. 2009. *The Passage to Cosmos: Alexander von Humboldt and the Shaping of America.* Chicago: University of Chicago Press.

Wigen, K. 2005. 'Discovering the Japanese Alps: Meiji Mountaineering and the Quest for Geographical Enlightenment'. *Journal of Japanese Studies* **31**(1): 1–26.

Wallace, A.R. 1869. *The Malay Archipelago.* New York: Harper.

— 1876. *The Geographical Distribution of Animals.* New York: Harper.

— 1880. *Island Life.* London.

Author Biographies

Joseph Christensen is a Postdoctoral Fellow at the Asia Research Centre, Murdoch University, Western Australia. He is the co-editor of *Historical Perspectives of Fisheries Exploitation in the Indo-Pacific* (2014) and of a forthcoming volume on natural hazards and peoples in the Indian Ocean World.

Christian Fleury has a Ph.D. in Geography on the subject of 'Island borders'. He is an Associate Researcher at ESO Caen, a research centre in Social Geography based at the University of Caen Basse-Normandie. A member of the Editorial Board of *Shima: The International Journal of Research into Island Cultures*, his interests are centred on three fields of study: sea appropriation conflicts, island issues and border effects.

John R. Gillis is Professor Emeritus of History at Rutgers University in New Jersey, USA, now dividing his time between summers on a small island in Maine and a permanent residence in Berkeley, California. He is author of *Islands of the Mind* (2004) and *The Human Shore: Seacoasts in History* (2014). He intends to keep working at the margins, where the most fascinating things are always to be found.

Glenn M. Grasso earned his Ph.D. in 2009 at the University of New Hampshire. He has taught at UNH and the United States Coast Guard Academy, and is currently lead historian for a historic document dealer in New York. Grasso has previously published on nineteenth century sailor, author and musicologist Frederick Pease Harlow, as well as on the rise and fall of the nineteenth century North Atlantic halibut fishery. His dissertation, *The Maritime Revival: Antimodernism, Class, and Culture, 1870-1940*, explored how the Anglo-American world re-imagined maritime historical events to serve contemporary political and cultural ends.

Stefan Helmreich is professor of anthropology at MIT. He is the author of *Alien Ocean: Anthropological Voyages in Microbial Seas* (University of California Press, 2009).

Adam Keul is a cultural geographer at the University of Connecticut's Avery Point campus. His work addresses experiences and political economies of touristed spaces. He has researched the production and performance of space relative to swamps in the southern US, the globalised Arctic and coastal New England.

Petr Kopecký is Associate Dean for International Relations and Development at the Faculty of Arts, University of Ostrava, Czech Republic. His first book is titled *The California Crucible: Literary Harbingers of Deep Ecology* (University of Ostrava Press, 2007). Its updated edition was published in 2013 (Lambert Academic Publishing). In 2003–2004 Kopecký was a Fulbright research scholar at San José State University, California. In 2013, he published a book entitled *Robinson Jeffers a John Steinbeck: Vzdálení i blízcí* in which he examines the contact zones in the writings of these two Californian authors through an ecocritical perspective.

Alexander Kraus is a Teaching and Research Assistant at the Department of History at the University of Münster. There he established the *Schreiblabor Geschichte* (history writing lab), realising public history projects with his students. The latest outcomes have been the books *Eine Geschichte Münsters in 30 Objekten* (2014, Aschendorff Verlag), *Stadt auf Kohle. Ein Lesebuch zur Geschichte der Zechen in Gelsenkirchen* (2014, Klartext) and *Umkämpfte Spielfelder. Fußball in der Geschichte Lateinamerikas* (2014, MV-Wissenschaft). He holds a Ph.D. in Eastern European History, switched to the history of science and is currently working on a sound story of the oceans from 1800 until today. He recently co-authored the book *Weltmeere. Wissen und Wahrnehmung im langen 19. Jahrhundert* (2014, Vandenhoeck & Ruprecht).

Matthew McKenzie is Associate Professor of History and American Studies at the University of Connecticut. He took his Ph.D. in Maritime History from the University of New Hampshire in 2003. From 2003–2006, McKenzie sailed offshore in the Atlantic and Pacific Oceans, teaching Maritime Studies with the Sea Education Association in Woods Hole, Mass., and filling-in as Assistant Engineer, deckhand and science deck lackey. He came to UConn's Avery Point campus in August 2006 and completed his first book, *Clearing the Coastline* (University Press of New England, 2011), exploring perceived changes in nineteenth century southern New England's inshore marine ecology. His current project, Breaking the Banks, explores the twentieth century intersection of popular representation, science and fisheries management in the chronic failure of the New England cod fishery. He currently represents the State of Connecticut on the New England

Fishery Management Council, the federal body tasked with managing regional marine fisheries.

Stuart J.L. Morrison is an interdisciplinary environmental historian from Stirling, Scotland. His doctoral thesis, *Vestfirðir, and the Emergence of Specialist Fishing Communities in Iceland* was defended in 2012 at the University of Stirling. He has written various articles on the environmental history and geoarchaeology of maritime and terrestrial environments in Scotland and Iceland, and is currently within the alcohol industry working with his other passion – Scotch whisky.

Christopher L. Pastore is Assistant Professor of History at the University at Albany, State University of New York. He holds a Ph.D. in American History and M.S. in college teaching from the University of New Hampshire as well as a B.A. in Biology from Bowdoin College and M.F.A. in nonfiction Creative Writing from the New School for Social Research. He is the author of *Between Land and Sea: The Atlantic Coast and the Transformation of New England* (Cambridge, MA: Harvard University Press, 2014), which examines the Narragansett Bay (Rhode Island) watershed from first European settlement to the early nineteenth century.

Michael S. Reidy is Professor of History at Montana State University. He is the author of *Tides of History* (Chicago, 2008) and co-author of *Exploration and Science* (ABC-Clio, 2007) and *Communicating Science* (Oxford, 2000). He recently co-edited *The Age of Scientific Naturalism: Tyndall and His Contemporaries* (Pickering & Chatto, 2014) and is the co-general editor of the Tyndall Correspondence Project. His current research focuses on how mountaineering changed science in the nineteenth century.

Robert Snyder is President of the U.S.-based Island Institute. As the head of the Island Institute, Dr. Snyder is responsible for working with island and remote coastal leaders to identify innovative approaches to community sustainability. Dr. Snyder received his Ph.D. from York University in cultural anthropology with an emphasis on the cultural politics of development.

Alistair Sponsel is Assistant Professor of History at Vanderbilt University. His research focuses on science and exploration.

Franziska Torma studied history and theatre studies in Munich and Bochum. In 2009 she completed her doctoral thesis, in which she focused on German scientific expeditions that described and mapped Central Asia between 1890 and 1930. She was research fellow at the Rachel Carson Center for Environment and Society, Munich, from 2009 to 2012 and John F. Kennedy Memorial Fellow at the Minda de Gunzburg Center for European Studies, Harvard University from 2012 to 2013. Since 2012 she has been Wissenschaftliche Mitarbeiterin (Assistant Professor) at the Chair for European Cultural History, University of Augsburg. Franziska has published on the history of mountaineering, animal protection issues in Africa and colonialism, with special reference to Germany's colonial culture and ideology. Her research interests include the history of science, the cultural and environmental history of the nineteenth and twentieth century, postcolonial studies and approaches of the spatial turn. Her current research project is entitled 'Germany's Seven Seas: Marine Biology and Environmental Globalism during the Long Twentieth Century'. She is also working on a textbook entitled *Umweltgeschichte nach 1945* (Environmental History after 1945).

Index

Index

D

Index

Index

N

O

P

Index

T

Index